**Astrophysics and Space Science Proceedings**

Volume 51

More information about this series at http://www.springer.com/series/7395

Gabriele Gionti, S.J.
Jean-Baptiste Kikwaya Eluo, S.J.
Editors

# The Vatican Observatory, Castel Gandolfo: 80th Anniversary Celebration

Springer

*Editors*
Gabriele Gionti, S.J.
Vatican Observatory
Rome
Holy See (Vatican City State)

Jean-Baptiste Kikwaya Eluo, S.J.
Vatican Observatory
Rome
Holy See (Vatican City State)

ISSN 1570-6591 ISSN 1570-6605 (electronic)
Astrophysics and Space Science Proceedings
ISBN 978-3-319-67204-5 ISBN 978-3-319-67205-2 (eBook)
https://doi.org/10.1007/978-3-319-67205-2

Library of Congress Control Number: 2017952501



Printed on acid-free paper

This Springer imprint is published by Springer Nature
The registered company is Springer International Publishing AG
The registered company address is: Gewerbestrasse 11, 6330 Cham, Switzerland

# Foreword

The history of the universe is made up of a web of stories. Some are cosmic or biological, and others are institutional and personal stories. All of this is the history of salvation that for us believers has Christ, the Crucified and Risen Lord, as the beginning and end of the entire universe.

Within this Big History, the history of the Vatican Observatory is inscribed and made of the personal stories of its members. One morning, in one of our daily meetings dedicated to discuss the usual business of the Observatory, Fr. Jozef Maj, S.J., vice-director for administration, proposed to bring together all members of the staff, including those located at the headquarters of Castel Gandolfo and at the Vatican Observatory Research Group in Tucson. His suggestion also included the adjunct scholars of the Observatory. I liked his proposal very much and decided to call for a symposium. It should be said that bringing together all the staff members and adjunct scholars was not an easy task; however, the occasion was worth the effort we made. And so, during the week of September 14–18, 2015, we celebrated the eightieth anniversary of the reestablishment of the Observatory at Castel Gandolfo by Pius XI, who also decreed that its management be entrusted to the Society of Jesus. On September 18, we concluded the celebration with the joy of being received by Pope Francis in private audience just on the eve of his trip to the USA. We appreciated very much the kindness and generosity of Pope Francis for receiving on such a busy day.

This volume gathers the proceedings of the symposium by the staff members and adjunct scholars. It covers a variety of topics ranging across the different fields of research in astrophysics, as well as the history of Astronomy, interdisciplinary studies, and the challenges that the mission of the Observatory faces in the dialogue between science and faith.

The Big History, as well as the history of the Vatican Observatory, has a promising future because it is filled with God's surprises. I wish the best to the current director, Brother Guy Consolmagno, S.J., and to the members of the Observatory. May they fulfill their mission with peace and joy.

I conclude on a personal note. The papal audience was my last act of director. It was a wonderful way to end my nine years' tenure. It was a privilege and an honor to serve the Holy See at the Vatican Observatory for 15 years. I take this opportunity to deeply thank Benedict XVI and Francis, who visited the Observatory showing their interest and support to the work of the Observatory. I am also very grateful to the Vatican authorities of the Governorate, the Secretariat of State and other Vatican offices with which I could collaborate, and to the Superiors of the Society of Jesus who supported my work and provided fine men to the staff of this research institute. I am profoundly grateful to the Jesuits and employees for their understanding and support. I am also thankful to the benefactors, friends, and collaborators who support the Observatory's mission in so many different ways.

Córdoba, Argentina
June 2017

José G. Funes, S.J.
Former director of the Vatican Observatory

# Address of His Holiness Pope Francis

The Holy See

## To Participants in the Symposium Organized by the Vatican Observatory

**Hall of Popes**

**Friday, September 18, 2015**

Dear Brothers and Sisters,

I welcome all of you who represent the working community of the Vatican Observatory, and I thank Cardinal Giuseppe Bertello for introducing our meeting.

"Deum Creatorem venite adoremus." With these words engraved on a marble plaque on the wall of one of the telescope domes at the Papal Residence in Castel Gandolfo, Pius XI began his discourse inaugurating the new Observatory on September 29, 1935.

Indeed, rather than a scientific problem to be solved, the universe is a joyful mystery to be contemplated with gladness and praise (cf. Encyclical Laudato Si', n. 12). "The entire material universe speaks of God's love, his boundless affection for us" (ibid., n. 84). St Ignatius of Loyola truly understood this language. He recounted that his greatest consolation was looking at the sky and the stars because this made him feel a tremendous desire to serve the Lord (Autobiography, 11).

With the reestablishment of the Observatory at Castel Gandolfo, Pius XI also decreed that its management be entrusted to the Society of Jesus. Throughout these years, the Observatory astronomers have continued the paths of their research, creative paths, following in the footsteps of the Jesuit mathematicians and astronomers of the Collegio Romano, from Fr Christoph Clavius to Fr Angelo Secchi, to Fr Matteo Ricci, and so many others. On this anniversary, I am also pleased to recall the discourse that Benedict XVI addressed to the Fathers at the last General Congregation of the Society of Jesus, in which he stated that the Church has a pressing need of religious ready to dedicate their life to being on the very frontier between faith and human knowledge, faith and modern science.

In these days, you, Fathers and Brothers, along with scholar associates, gathered to discuss your research and topics concerning the dialogue between science and religion. In this regard, St John Paul II stated: "What is important is that the dialogue should continue and grow in depth and scope" (Letter to Fr George V. Coyne, SJ, 1 June 1988). And he asked: "Is the community of the world religions, including the Church, ready to enter into a more thoroughgoing dialogue with the scientific community, a dialogue in which the integrity of both religion and science is supported and the advance of each is fostered?" (cf. ibid.).

Within the context of interreligious dialogue, now more than ever, scientific research on the universe can offer a unique perspective, shared by believers and non-believers, which can help to achieve a deeper religious understanding of creation. In this sense, the Schools of Astrophysics, which the Observatory has organized in the last 30 years, present a valuable opportunity for young astrophysicists from around the world to dialogue and cooperate in the search for the truth.

Additionally during your conference, you also discussed the importance of communicating that the Church and her Pastors foster, encourage, and promote authentic science, as Leo XIII emphasized (cf. Motu Proprio Ut Mysticam). It is very important that you share with people the gift of your scientific knowledge of the universe, freely giving what you have freely received.

In the spirit of gratitude to the Lord for the witness of science and faith that the members of the Observatory have borne in these decades, I would like to encourage you to continue the journey with your colleagues and with those who share the enthusiasm and toil of the exploration of the universe. It is a journey that you also take in the company of the Observatory staff, of benefactors and friends, and of so many people of good will. Yes, we are all on a journey to the common house of heaven, where we will be able to read with admiration and happiness the mystery of the universe (cf. Encyclical Laudato Si', n. 243).

May Almighty God, who sustains in existence the whole universe, through the intercession of the Virgin Mother, fill you with his peace and bless you.

Vatican City, Vatican City State Pope Francis

# Contents

# Introduction

On September 29, 1935, Pope Pius XI officially inaugurated the new headquarters of the Specola Vaticana in the Papal Palace of Castel Gandolfo. With new telescopes, a new spectra laboratory, and a young staff of Jesuit scientists, this inauguration marked the beginning of an intense period of scientific achievement at the Vatican Observatory.

From September 14 to 18, 2015, eighty years later, the current members of the Vatican Observatory and its adjunct scholars participated in an internal symposium to celebrate the anniversary of this move from Vatican City to Castel Gandolfo. This was also an occasion to celebrate the new headquarters of the Vatican Observatory in Castel Gandolfo; in 2009, the Vatican Observatory headquarters moved from the Papal Palace to the buildings that once belonged to the convent of a community of Basilian nuns, expelled from White Russia, and given asylum in the Vatican under direct protection of Pius IX.

The roots of the Vatican Observatory date back to the reform of the Calendar, 1582, when Pope Gregory XIII promoted the reform of the Julian Calendar into a new and more precise version, called the Gregorian Calendar, which is, in fact, the one that we all use and can find hanging in our houses and offices today, particularly in the Western world. One of the main reasons for this change was a new determination of the date of Easter and other religious feasts. The new calendar was the result of a study commissioned to a group of scientists that included the Jesuit father Cristopher Clavius, S.J., who wrote a book on the new Gregorian calendar for the purpose of spreading its use. This reform involved a series of astronomical observations that were demonstrated in the Tower of Winds in Vatican City, considered by some to be the first Vatican Observatory. Pope Leo XIII in the Motu Proprio "*Ut Mysticam*" (1891), the foundation act of the Vatican Observatory, explicitly referred to it as the first Vatican Observatory, where Fr. Ignazio Danti, O.P. (Dominican), built a meridian line. Traces of a Vatican Observatory continued to exist in Vatican City. A new "Specola Vaticana" (Vatican Observatory) was opened in the Tower of Winds by Mons. Filippo Luigi Gilii in 1789 and lasted until 1820.

The legacy of the Vatican Observatory also includes the Pontifical Observatory of the Roman College, located on the roof of Saint Ignatius Church in Rome. Here, the most important figure was Fr. Angelo Secchi, S.J., who is considered to be the father of modern astrophysics; he was the first person to propose a star classification based on the stars' spectra. After the fall of the Pontifical States in 1870, following the entrance of the Italian troops and the relocation of the capitol of the newly formed Italian state to Rome, the Holy See lost all of its Observatories (though Fr. Secchi was allowed to continue his work at Saint Ignatius until his death in 1878). Therefore, under the suggestion of Fr. Francesco Denza, a Barnabite father who was a student of Fr. Secchi, Pope Leo XIII refounded the Vatican Observatory in Vatican City in the Tower of the Winds and appointed Fr. Denza as director. From its beginnings, the newly refounded Vatican Observatory was involved in very important projects, most famously the Observatory's participation alongside many other national observatories in the Carte du Ciel project, the first photographic mapping of the sky; the Vatican Observatory obtained a special "Carte du Ciel" telescope for this purpose. Succeeding Denza, the newly appointed director, Fr. Rodriguez, was a meteorologist, not an astronomer; questions raised by his leadership forced his rapid replacement. Pius X instituted a committee headed by Card. Pietro Maffi to find a new director. Fr. Hagen, S.J., director of the Observatory of the Jesuit University of Georgetown, was then appointed the new director of the Vatican Observatory in 1906. Since then, the Vatican Observatory has been entrusted to the Society of Jesus, with a community of Jesuits established to staff the Observatory, in both scientific and administrative capacities, on the model of the magazine "La Civiltà Cattolica."

Eventually, light pollution, caused by the increasing number of artificial lights in Rome, ruined any chance of observing the sky in the city. Therefore, as already mentioned, the headquarters of the Vatican Observatory moved from Vatican City to Castel Gandolfo in 1935, along with an augmented staff of young scientists, leading to a robust increase in the Specola's scientific activities. But the problem of light pollution never really stopped, and even though a one-meter Schmidt telescope was donated by Pius XII to the Specola, in the 1980s, the skies above Castel Gandolfo were already so light-polluted as to hamper any kind of scientific observation. A further migration was necessary. Many proposals were examined; finally, Fr. George Coyne, S.J., who had been appointed director in 1978, established a new branch of the Vatican Observatory in Tucson, Arizona. The offices of the astronomers of the Vatican Observatory were located in the Astronomy Department of the University of Arizona, The Steward Observatory, thus enabling them to use the telescopes and other facilities of the university. A few years after their arrival in Tucson, the Vatican Observatory accepted an offer to build a new telescope along with the Steward Observatory on the top of a 3000-m mountain, Mt. Graham.

The research carried out at the Vatican Observatory covers a broader range than that started 80 years earlier in Castel Gandolfo. This is in part due to the different fields of expertise of the current staff members, but it also reflects the different research interests in astronomy today. For example, the papers presented in these proceedings highlight a conspicuous group of people, quite numerous given the size of the Vatican Observatory, doing research in Planetology. The paper on the meteorite collection of the Vatican Observatory presents a history of the collection and describes the actual research on meteorites: measurements of heat capacity, porosity, and other physical properties. Near-Earth Object (NEO) is also an active field of study. Here, more than a mere astrometrical study of meteors, the main focus is the study of the physical properties of the NEO through spectroscopic analysis of the light that comes from them.

An equally active field of research at the Vatican Observatory is stars and stellar evolution. One field of research is the λ-Bootis stars, which appear to be quite enigmatic for their classification based on spectroscopic methods. The relative abundance of their metallic elements especially poses some serious questions not yet resolved. Researches in this direction might give interesting information about the structure and evolution of certain stars in our Milky Way, as well as improving the models we use to describe the evolution of stars. The VVV (Vista Variable in the Via Lactea) survey is instead a different kind of research, which aims to map the bulge of our Galaxy through the 4-m Vista telescope of the European Southern Telescope (ESO). The aim of this research is to get not only a better picture of the bulge of our Galaxy but also a better picture of its interior part. Of course, in this way, it is also possible to improve the knowledge one has of our Galaxy. An interesting field of research is also being performed in the field of stellar evolution. Here, the field had a milestone in the joint Vatican Observatory–Pontifical Academy of Science conference on Stellar Population in 1957. Researches in this direction are being continued in the study of stellar evolution. In particular, through numerical simulations, the metallicity of host subdwarf stars is being studied, since this can shed light on the general picture of star evolution. Another quite established line of research at the Vatican Observatory is the use of the Vilnius photometric system. It is a medium-band seven-color photometric system, created in 1963 for the classification of stars. Originally, it was applied to photographic plates, but lately, it has also been used for CCD cameras.

The study of Galaxies has been an important topic of research at the Vatican Observatory over the last twenty years. A promising line of research is the study performed by the Wide Field Nearby Galaxy Survey (WINGS) group. This study helps in understanding the formation and evolution of galaxies. Jellyfish Galaxies are also studied. They have been stripped of their gas and have a tentacle of debris, which is the main reason for their name. Their study is useful for understanding Galaxy formation and evolution, as well as the formation and evolution of stars. The study of Galaxy halos has been another interesting topic of discussion at the meeting. Galaxy halos are particularly useful for detecting dark matter. In fact, these are the "classical" regions in which dark matter is generally located.

Cosmology has been another subject of discussion at this conference. There has been interesting talk about non-commutative geometry, in which the topic of dark matter and dark energy has been tackled. It has been shown that, using techniques of non-commutative geometry, it is possible to derive a version of the Einstein's Equations with additional terms that can be added to the energy–momentum tensor and can then be interpreted as additional matter. Another topic treated has been the study of CMB radiation. The Standard Cosmological Model (usually called ΛCDM, in which Λ means the cosmological constant and then the presence of dark energy, and CDM stands for cold dark matter) can be extended with additional terms. Adding these terms into the theory, the possible effects that they have on CMB radiation can be calculated, and then, it could be possible, in principle, to check directly whether this modification of the Standard Cosmological Model really does exist. A final cosmological paper discussed at the conference has been on the issue of dualities in cosmology. It has been presented that in alternative theories of gravity (also called F(R) theories of gravity), the FLRW (Friedmann, Lemaitre, Robertson, Walker) homogeneous and isotropic metric, which is the solution of the Standard Model in Cosmology, is also the solution of a general extended theory of gravity. The cosmological model exhibits further symmetries, which are duality transformations.

A section of the meeting has been dedicated to the history of Astronomy. There has been a presentation on the history of the Vatican Observatory, the content of which was already mentioned in the first part of this introduction. A second paper has been on the recent finding of quite an old telescope lens stored in Monte Mario. This telescope, the Cauchoix, has important historical relevance, since it was bought by the Astronomical Observatory of the Roman College and was used by Fr. Angelo Secchi, S.J., for the first stellar classification based on light spectra.

A final section of the conference has been focused on the dialogue among science, theology, and philosophy. In a theology paper, the issue of scientism has been discussed very carefully, highlighting that the usual *clichè* on the dominance of scientific culture in modern society should be carefully rethought. Some suggestions are given to believer scientists in order to promote a new evangelization. In a philosophy paper, it has been discussed what a philosophical discussion is, and how hard it is nowadays to carry on such a discussion on topics concerning cutting-edge research in the natural sciences. It is analyzed that these disciplines also have to appeal to an authority, when the validity of a scientific theory needs to be accepted. Progress in philosophical reflections on the natural sciences is envisaged in interdisciplinary analysis, and the Vatican Astronomers, because of their education, are seen as having the skills to perform them. A final paper in the conference has dealt with the topic of scientific theory being theory-laden. Some philosophers of science have pointed out that the majority of scientific observations are influenced by the cultural environment in which they are made. Scientists seem to observe nature through eyes that are influenced by the same theory they are trying to build. This, fostered to extreme consequences, appears to create a strong dependence for the natural sciences on the social context, weakening the presumption of objectivity that is usually attached to scientific investigations. The Astronomers of the Vatican Observatory, under the force of their training in both the humanities and spirituality, are seen as people who are key to seeing these aspects of natural sciences and reflecting on them critically. They could go a long way towards regaining the credibility of science, which has been lost in this time of postmodernism.

This work is dedicated to the loving memories of Bill Stoeger, S.J. and Luiji Lozi.

Participants of the 80th anniversary celebration with Pope Francis on September 18th, 2015

Pictures courtesy of the mail post office of the Vatican City State

# Part I
# Planetary Science

# 108 Years of Meteorites at the Vatican Observatory

**Robert J. Macke**

**Abstract** The Vatican Observatory today hosts approximately 1200 meteorite specimens of all major meteorite types from around the world. The collection originated in a series of donations from Adrien Charles Marquis de Maurois, with the majority of specimens donated after his death by his widow in 1935. The collection has grown slowly since then, primarily through trades and donations. The collection boasts a number of historically and scientifically significant specimens, including a piece of the Martian meteorite Nakhla that has been handled by popes. The collection has been used for research in various forms since its early days. In the 1930s, it served as an early test bed for spectroscopic studies with the intention of comparison with asteroid spectra. In the 1990s, the collection served as a proving ground for pioneering work on meteorite porosity measurement using ideal-gas pycnometry and glass-bead immersion, which has become a widely adopted technique. In this past decade, it again served as a proving ground for new techniques; this time for whole-stone low-temperature heat capacity measurement by liquid nitrogen immersion.

## Introduction

The astronomers at the Vatican Observatory study everything from cosmology, theoretical physics, stars and galaxies to the Sun, planets, and asteroids. To do this, they make use of theoretical calculations, computer simulations, and telescopic and other observations. In all of these cases, astronomers are physically separated from the object of their study by vast expanses of empty space and must rely on the data provided by light that has traveled across the void to reach the Earth.

---

The title is taken from the presentation given at the conference in commemoration of 80 years of the Vatican Observatory in Castel Gandolfo, which took place in September 2015.

---

R. J. Macke (✉)
Vatican Observatory, 00120 Castel Gandolfo, Vatican City State, Italy
e-mail: rmacke@specola.va



G. Gionti, S.J. and J.-B. Kikwaya Eluo, S.J. (eds.), *The Vatican Observatory, Castel Gandolfo: 80th Anniversary Celebration*, Astrophysics and Space Science Proceedings 51, https://doi.org/10.1007/978-3-319-67205-2_1

In the planetary sciences, research into the history and origins of the solar system offers two further approaches that bring the solar system closer to home: spacecraft missions and sample-return missions. (The astronomers of the Vatican Observatory are not directly involved in these missions, but we have collaborated with scientists who are.) Spacecraft can bring instrument packages very close to planets, moons, and other objects of interest and send large amounts of photos and data back to the Earth for study. In some cases, the spacecraft can retrieve some material and bring these samples of other solar system bodies back to the Earth to study in the lab. Because of the expense and technical challenge involved, these sample-return missions have so far been limited to the manned Apollo and unmanned Russian Luna missions to the Moon, and Genesis, Stardust, and Hayabusa, which each retrieved small quantities of solar wind particles, comet dust, and asteroid dust, respectively.

Fortunately, nature provides its own very low-budget sample return missions in the form of meteorites. Meteorites are rocks that originated on other bodies in the solar system and that have fallen from space to the surface of the Earth. They can then be collected and brought to museums and laboratory facilities where they can be studied in comfort. One such facility is housed at the Vatican Observatory and hosts the Vatican meteorite collection. It boasts over 1200 individual specimens representing more than 500 distinct meteorites (Fig. 1).

Meteorites predominantly come from asteroids, but it is possible to get material from other solar system bodies. A limited number of meteorites have been determined to have originated on either the Moon or Mars. In these cases, they were knocked off the surface by an impact with enough energy that they escaped the gravity well and could float freely in space for a time until, eventually, they were captured by the gravity well of the Earth and fell down. Despite many space missions to the surface of the red planet, the only Martian material we have available for laboratory study is in the form of these meteorites.

Of the asteroidal meteorites, they originally lithified in larger parent bodies that subsequently broke up in the early solar system to form the asteroids as we know them today. Some originated in parent bodies that were large enough to melt and differentiate into a core, mantle and rocky silicate crust. Meteorites that originate from such bodies include iron meteorites (from the cores), stony achondrites (from the crust), pallasites and mesosiderites (from somewhere deep in the interior). Some —the chondrites—originated in parent bodies that did not differentiate and so still preserve signatures of the conditions of their formation and the formation of the solar system itself. They get their name from the presence of chondrules, which are solidified silicate melt droplets that formed in the early solar nebula, under conditions that are still not well understood but predate the formation of the parent body. These objects provide a lot of material for study into the formation and history of the solar system.

**Fig. 1** The Vatican meteorite collection (top) the display cabinet (bottom) the remainder of the collection laid out on the bench tops of the meteorite lab

## *Adrien Charles Marquis de Mauroy (1848–1927)*

The Vatican meteorite collection began through the donation of a few meteorites by the French Marquis of Mauroy, Adrien Charles (Fig. 2). He was a member and one-time president of the French Mineralogical Society, and was an avid collector of

**Fig. 2** Adrien Charles Marquis de Mauroy (1848–1927)

meteorites and mineral specimens around the turn of the 20th century. At its peak, his was one of the largest private meteorite collections in the world (Consolmagno 2006; Maffeo 2001). When Adrien-Charles was forming his collection, most meteorites were only identified by being observed falls. As a result, most of the meteorites in his collection come from historically significant events. One example is the meteorite Ensisheim, of which he had several specimens. This meteorite fell in 1492 over the town of Ensisheim in what is today France, and is one of the earliest well-documented meteorite falls in Europe (Fig. 3). The meteorite L'Aigle, which also fell over France in 1803, was documented for the French Academy of Sciences by Jean-Baptiste Biot (also famous for the Biot-Savart law of electromagnetics), and established in the scientific community that these objects did indeed fall from the sky, which was a matter of some controversy at the time. The meteorite Weston fell over Connecticut a couple years later, establishing the same fact within the American scientific community. Of this event, Thomas Jefferson is apocryphally attributed with commenting that he would rather believe that two Yankee scientists would lie about it than believe that rocks would fall from the heavens.

The marquis was also a good Catholic and friend of the Church. In 1896, he proposed the founding of a natural history museum in the Vatican, with the promise

**Fig. 3** A scene from Sigismondo Tizio's (1528) account the fall of the Enshisheim meteorite in 1492, reproduced from a woodcut broadsheet by Sebastian Brandt (1492) (*Photo credit* Biblioteca Apostolica Vaticana)

of donating a large portion of his meteorite and mineral collection. The intent was to place the museum in the same facility as the Vatican Observatory, but at the time there was not sufficient space where the observatory was housed within the Tower of the Winds. Therefore it was not until 1907, after the observatory expanded to include a facility in the Vatican gardens near the Leonine Tower, that he donated 104 meteorites, mostly duplicates of other specimens in his collection. He made another donation of 50 meteorites in 1912. In 1927, the marquis passed away, and several years later in 1935 (the year of the move of the observatory from within the walls of the Vatican to the Papal summer residence in Castel Gandolfo), his consort widow Marie Caroline Eugénie donated the majority of the remainder of his collection to the Vatican (Consolmagno 2006; Maffeo 2001).

In all, about 1000 meteorite specimens and almost as many terrestrial mineral specimens were donated. The terrestrial minerals were subsequently sent on open-ended loan to another institution in Austria, where unfortunately they became hopelessly intermixed with the local collection and are effectively lost to the Vatican. The meteorites, however, remain and form the core of the Vatican meteorite collection today. Unfortunately, the dream of founding a natural history museum was never completely realized, but the underlying intent to support the Vatican in its pursuit of science was, and the remainder of this chapter will discuss the scientific contribution made by these valuable meteorites.

## *Growing the Collection*

While the majority of the Vatican meteorite collection originated from the Marquis' generosity, the collection has grown slowly over the subsequent 80 years, through donations and the occasional trade. One notable example of this is the meteorite Nakhla, which fell over Egypt in 1911. John Ball of the Geological Survey of Egypt, hoping to establish a trading relationship with the Vatican Observatory, donated a piece weighing more than 150 g (Consolmagno 2006; Fig. 4a).

(a)

(b)

(c)

**Fig. 4** **a** The Nakhla meteorite; **b** examined by Pope Benedict XVI during his visit in 2009; **c** examined by Pope Francis during his visit in 2013 with then-curator Br. Consolmagno explaining the significance of the specimen (*Photo credits* for (**b**) and (**c**): L'Osservatore Romano)

The origin of this unusual meteorite was not known at the time, and it was not until after the Viking space missions of the 1970s that Nakhla and related meteorites were determined to have come from the planet Mars. The Vatican specimen of Nakhla is particularly unique because it has been handled by popes: Both Pope Benedict XVI and Pope Francis held the meteorite during their visits to the Observatory in 2009 and 2013, respectively (Fig. 4b, c). Another notable specimen is a small piece of lunar regolith collected in 1972 by astronauts Eugene Cernan and Harrison "Jack" Schmitt during the Apollo XVII mission and given by the United States as a diplomatic gift to the Vatican City-State, along with a small Vatican flag that flew on the mission (Fig. 5). (There is a similar moon rock from Apollo XI on

**Fig. 5** A specimen from the Apollo XVII mission given to the Vatican, along with a Vatican flag that flew on the mission. The text reads (above): "This fragment is a portion of a rock from the Taurus Littrow Valley of the Moon. It is given as a symbol of the unity of human endeavor and carries with it the hope of the American people for world peace." (below): "This flag of your nation was carried to the Moon aboard Spacecraft America during the Apollo XVII mission, December 7–19, 1972/Presented to the people of the State of the Vatican City/From the people of the United States of America/Richard Nixon/1973"

display in the Vatican Museums.) One of the most recent additions to the Vatican collection, donated in 2015 by a collector near Chicago, is a small piece of the Chelyabinsk meteorite. This came from an object that famously exploded over a crowded city in Russia in 2013, destroying windows for many miles around and causing a large number of injuries (Ruzicka et al. 2015).

Today the Vatican meteorite collection has over 1200 specimens of about 530 distinct meteorites, representing both falls and finds from every major meteorite type, collected from every continent except Antarctica. Unfortunately, due to the market demand for meteorites and the subsequent high prices, it is difficult to continue to grow the collection and we continue to rely on the generosity of donors.

## Scientific Work with the Meteorite Collection: The Early Years (1935–1950s)

In the planetary sciences, there are many things that scientists wish to understand better. Among these are questions of the composition and structure of asteroids and planetary bodies, and questions of how these objects formed and what happened over the four and a half billion years of their history. Ultimately, we wish to better understand the formation and history of the solar system itself. These are the sorts of questions that meteorites are well-suited to answer, since laboratory studies of the isotopes, elements, minerals, and overall geology of meteoritical specimens bear clues to all of these questions if only we know how to look.

In order to take what we have learned from meteorites and apply them to planetary bodies, however, we first must have a good idea of which objects or at least which types of objects the meteorites came from. Even early on, it was reasonable to assume that most meteorites came from asteroids. Nevertheless, there are many different types of meteorites and many different types of asteroids, marked by different albedos (reflective brightnesses) and colors.

In the early 20th century, the use of spectroscopy in astronomy was still fairly new. The technique had been developed for studying starlight by notable astrophysicists such as Angelo Secchi and Fraunhofer in the mid-19th century, but was only slowly gaining favor as a general approach to astronomy. Very little work had been done on studying the spectra of asteroids. By the time the Vatican Observatory transferred from its original home within the walls of Vatican City to the Papal summer residence at Castel Gandolfo in 1935, spectroscopy had become sufficiently important that it was seen to be a major part of the research program of the new facility. In the notice describing the decision of the Holy Father to move the observatory, then-director Fr. Johan Stein S.J. expressed his vision to extend the spectroscopy program to the study of meteorites. He writes: "Another point of the program will be the spectral examination of the meteorite collection of the Observatory, to elucidate not only the exact proportions of their component elements, but also, insofar as means permit, to research the more subtle structures of the cosmic elements and to compare the relative number of the isotopic components in terrestrial and cosmic bodies." (Stein 1933, *my translation*)

With this in mind, the new facility was equipped with a spectroscopy laboratory. Fr. Stein appointed Fr. Alois Gatterer to set up and equip the laboratory in a large room on the main floor of the building (Fig. 6). This work was aided by the technical expertise of Br. Karl Treusch, who prior to becoming a Jesuit had worked for the Zeiss company in Germany. Much of the research was done in conjunction with Fr. Joseph Junkes, who became director of the laboratory after the death of Gatterer in 1953 (Maffeo 2001). Some of the very first spectroscopic studies of meteorites were conducted in this facility on the Vatican meteorite collection.

Unfortunately, very little in the way of publishable results were produced from the spectroscopic study of meteorites. Gatterer and Junkes did manage to produce one paper in 1940 "Uber den Steinmeteoriten von Rio Negro" ("On the iron meteorite from Rio Negro") in which they analyzed the metal taken from the Rio Negro meteorite, as well as two other meteorites: Lantzenkirchen and Albareto. Their analysis was based on both spectroscopy and wet chemistry (in which samples of meteorite are subjected to various chemical reagents to determine their chemical makeup). Later, in 1952, Fr. Ernst Salpeter wrote up a study of the chlorine content of twenty meteorites.

In retrospect, the reason for the dearth of publishable results is fairly obvious. Several factors come into play. The first is that meteorites are not simple minerals,

**Fig. 6** The spectroscopy laboratory (*Source* Salpeter 1952)

but are assemblages of a wide variety of mineral components intermixed tightly together, with the individual components ranging in size from microns to centimeters. In order to perform the analyses given the equipment available at the time, several grams of material would have to be pulverized and vaporized. This would further intermix all of the chemical species, resulting in spectra that are a forest of overlapping lines, and making it prohibitively difficult to isolate more than a few specific chemical lines (Fig. 7).

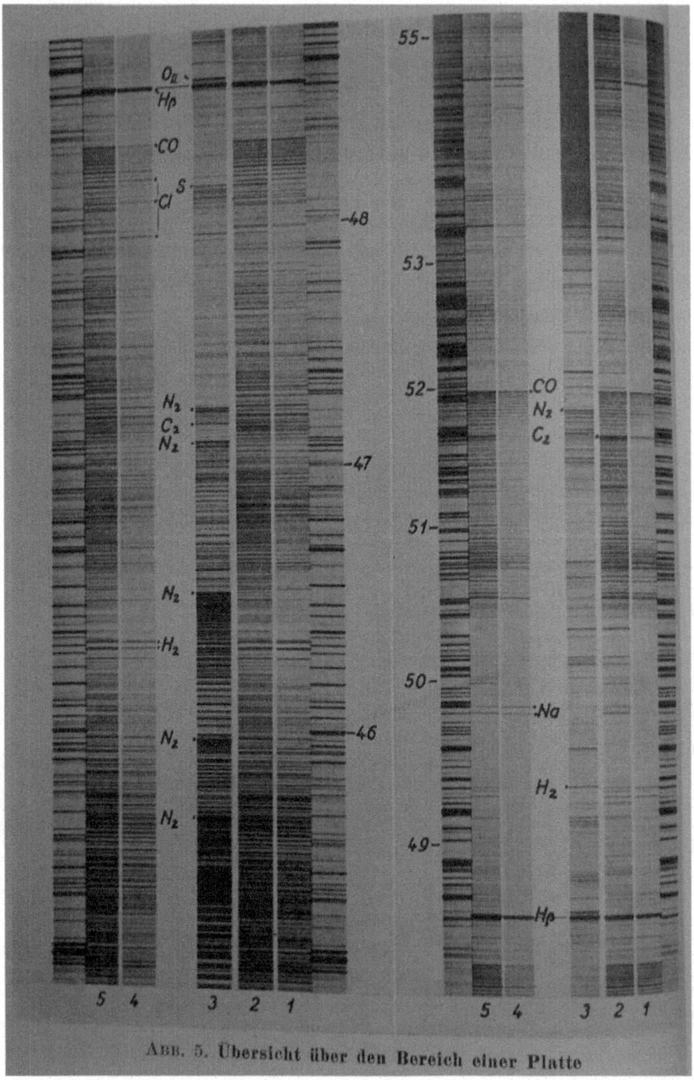


**Fig. 7** Spectral lines from Salpeter (1952). Using the photographic techniques of the day, specific lines are difficult to distinguish and analyze in crowded regions

Second, the interpretation of meteorite and asteroid spectra is still a difficult problem even today. Spectral studies of asteroids are complicated not only by the fact that the Earth's atmosphere is opaque at some important wavelengths, such as those corresponding to water, but also that the interaction of the surface of asteroids with solar wind particles, cosmic rays, micrometeorite impacts, and other things, alters their spectrum. Thus, even when there is a high degree of confidence that a certain meteorite type pairs with a certain asteroid type, their spectra do not look the same.

Furthermore, the early spectroscopic studies were limited by the available technology. Optical spectroscopy of whole rocks only provides limited data. Today, instruments such as scanning electron microscopes equipped with X-ray spectrometers are used to do quantitative chemical studies not only of whole meteorites, but of individual components. The isotopic studies that Stein foresaw are not at all possible with ordinary spectroscopy. Modern isotopic studies are performed using instruments such as secondary-ion mass spectroscopy, in which ionized atoms and molecule fragments are ablated off the surface of the sample and accelerated around a magnetic field that separates them by the ratio of charge to mass. Different isotopes of the same element are distinguished by their different masses.

Today, those kinds of instruments and their close relatives form the backbone of meteorite studies and not only tell us their composition, but also confirm their origin in solar system bodies beyond the Earth, and give us information about their ages. For instance, they tell us that some refractory inclusions, referred to as Calcium-Aluminum-rich Inclusions (CAIs) and found in very primitive meteorites, are among the very first materials to have solidified out of the early solar nebula over 4.56 billion years ago (cf. Marvin et al. 1970; Ireland et al. 1990). Ironically, due to the cost of purchasing and maintaining such equipment, the Vatican Observatory does not host any of the modern spectroscopic or other analytical equipment that is the industry standard for meteoritics today.

Though very little research was done on the spectra of meteorites themselves, the Vatican Observatory spectroscopy laboratory was indeed quite fruitful. Pure mineral species proved to be much easier to study, and filled an important void in existing research. Work done in the laboratory led to the founding by Fr. Gatterer and collaborators of the journal *Spectrochimica Acta* in 1939. Initially published in-house at the Vatican Observatory, this scientific journal became wildly successful, finally subdividing in 1967 into *Spectrochimica Acta A* (for molecules) and *Spectrochimica Acta B* (for atoms). Today, it continues to be published by Elsevier Press, which is one of the major publishers of scientific journals.

## Scientific Work with the Meteorite Collection: Recent Years (1993–2015)

### *Physical Properties*

With the exception of loans of specimens to researchers at other institutions, the meteorite collection remained largely unutilized from the late 1950 until the arrival in 1993 of Br. Guy Consolmagno S.J. He had been given the post of curator of the collection with the order to use the meteorites to "do good science" (Consolmagno, *personal communication*). With no direction beyond this, he had both the challenge and the freedom to develop a new research program. One of the problems he had to confront was the condition in which the meteorites had been curated over the past hundred years or more. In that time, terrestrial contamination from the atmosphere, moisture, dirt, human fingerprint oils, and other sources had permeated the collection. While this was not a concern to researchers in the early days of meteoritics, much of modern research on meteorites relies on fresh, pristine samples in which small amounts of terrestrial contaminants may skew results considerably. Br. Consolmagno had to find a way to study the meteorites that would not matter if the meteorite had been contaminated, and that would not merely repeat past measurements but would confront important scientific questions relevant to today.

One area prime for a new research program was meteorite physical properties such as density and porosity. Porosity was becoming increasingly important as a parameter in models of asteroid structure and formation. Astronomical studies of asteroids were revealing some with extremely low densities—in order to interpret these data, one needed to know the density and porosity of analogous materials (i.e., meteorites).

In order to determine porosity, two separate density measurements are necessary: bulk density and grain density. In general, density is the mass divided by the volume of the sample. Bulk and grain densities consider different volume contributions. Bulk density ($\rho_b$) relies on the entire volume enclosed by the outer envelope of the meteorite surface. It includes not only rocky material, but any volume contributions from pore space contained within the meteorite. Grain density ($\rho_g$), on the other hand, takes into account only the volume filled by solid matter, omitting any pore space. It is effectively the weighted average density of all of the mineral species present in the meteorite. Porosity ($P$) is calculated from the ratio of these two densities: $P = 1 - (\rho_b/\rho_g)$.

Early measurements of meteorite density and porosity at other institutions relied on techniques such as fluid immersion in either water or chemicals such as toluene (cf. Keil 1962; Pesonen et al. 1993; Fujii et al. 1981). These techniques inevitably led to further contamination of the samples, thus making them inapplicable to large numbers of meteorites. In addition, they were not necessarily reliable. Fluid incompletely flows into pore space, yielding results that are neither accurate for bulk density nor for grain density, and leaving all early porosity measurements questionable. Other techniques had also been tried, such as cutting samples into parallelepipeds (Yomogida and Matsui 1981, 1982, 1983) or wrapping the specimens in clay (Matsui et al. 1980), but these also had drawbacks in contamination or reliability and could not be widely applied.

For grain density, another option presented itself, which had been applied to a limited degree on other extraterrestrial materials, including a few Apollo moon rocks (Cadenhead et al. 1972; Cadenhead and Stetter 1975): ideal-gas pycnometry. This non-contaminating and non-destructive technique can be applied on whole stones of many grams in size. It utilizes two sealed chambers of known volume, connected by a valve. The first chamber, into which the meteorite is inserted, is pressurized with an inert gas such as helium that behaves well as an ideal gas. After the pressure is measured, the valve is opened to allow the gas to expand into the secondary chamber. As the volume expands, the pressure drops. By comparing the initial pressure with the final pressure after expansion, the volume displaced by the meteorite is calculated. Since the gas penetrates all pore space quite well in a typical meteorite, the volume produced is that of the solid component, i.e. the grain volume.

In 1997, through connections with his friend and collaborator Dan Britt, Br. Consolmagno acquired the use of an apparatus for ideal-gas pycnometry. This purpose-built device had been originally constructed by one of Britt's masters' students for measuring the density of drill cores (Geddis 1994; Fig. 8a), and so had a massive cylindrical primary chamber of 11.5 cm diameter and 30 cm in length, dwarfing just about every meteorite placed into it. Nevertheless, they were able to get some of the first reliable grain density measurements on several meteorites from the Vatican collection.

Measuring bulk density non-destructively and without further contamination presented a greater challenge for the two collaborators. A vital clue to their problem came during the morning coffee break that is part of the daily routine of the Observatory. Br. Consolmagno was pouring sugar into his cappuccino when he realized that the tiny grains were behaving collectively like a fluid as they flowed into his beverage. They got the idea of using small granular material to serve as an Archimedean fluid. Eliminating sugar, sand, and related grains as too unpredictable because of the sharp angular shape of the grains, they settled finally on small (initially, 40–80 μm in diameter, though later work used 700–800 μm) spherical glass beads (Fig. 9). In addition to being non-contaminating and well-behaved, they were also available in large quantities due to their industrial use in bead-blasting applications.

For this measurement, a meteorite is placed into a cup of known volume, which is then filled completely with beads. The beads flow around the meteorite and fill all the remaining volume. Taking the mass of the bead-filled cup, and knowing the average density of a quantity of beads (determined separately), the volume displaced by the meteorite can be calculated. Since the beads are larger than a typical surface crack, they do not penetrate into the interior and so this method produces a reliable bulk volume and density.

This technique, along with initial results for grain density, bulk density and porosity for 89 meteorites of various types were published in 1998 in a paper entitled "The density and porosity of meteorites from the Vatican collection." (Consolmagno and Britt 1998).

Around 2004, Consolmagno acquired through Dan Britt a new ideal-gas pycnometer, commercially produced by the Quantachrome corporation in Florida (Fig. 8b). This device, developed initially for studying the densities of powders, proved almost perfect for the measurement of hand-sized meteorites. Its primary chamber of 4.8 cm diameter by 7 cm in depth could comfortably fit most of the hand-sized stones in the Vatican collection, and it had volume-filling inserts for

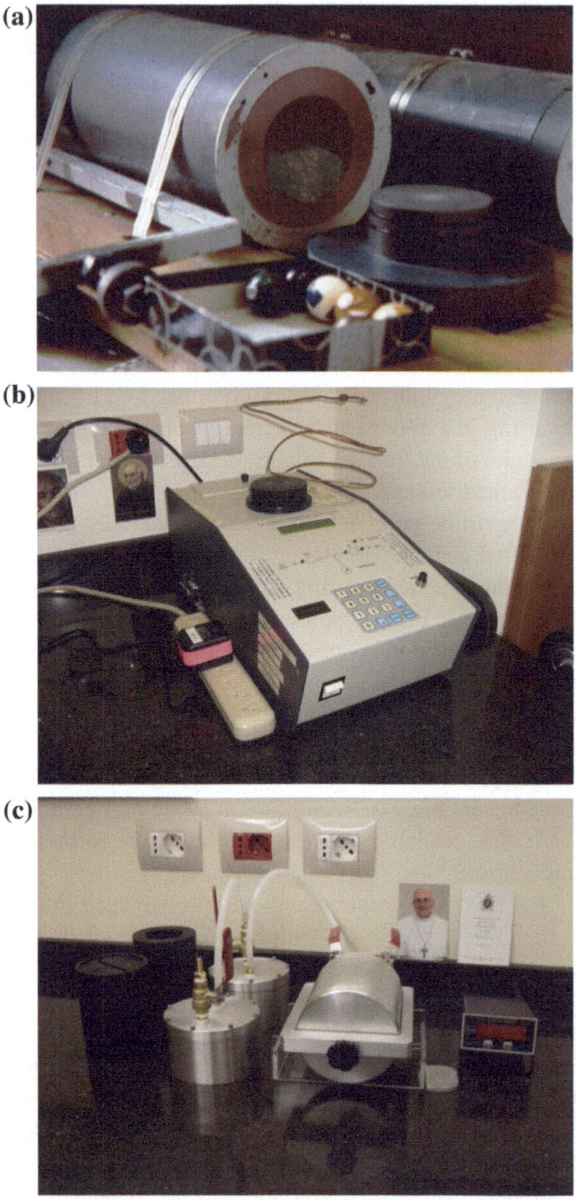

**Fig. 8** Ideal-gas pycnometers used at the Vatican Observatory. **a** Pycnometer constructed by Geddis and used for the earliest work in the 1990s. **b** Quantachrome Pycnometer 1000 in use at the Specola since 2004. **c** custom-built ideal-gas pycnometer designed by Br. Macke and in use at the Specola since 2013

even smaller samples. Since the uncertainty of measurement depends in part on the volume of the chamber relative to the size of the meteorite, these smaller chambers could produce much more precise and reliable results than the old device.

In the summer of 2004, Br. Consolmagno invited Br. Robert Macke, then studying philosophy in St. Louis, to spend a few months assisting with the measurements. Br. Macke gained proficiency in the use of the bead method, and

(a)
(b)

**Fig. 9** Glass bead apparatus. **a** Early apparatus using 40–80 μm beads. **b** Later apparatus using 700–800 μm beads (shown in use at NASA Johnson Space Center)

together the two determined densities and porosities for the majority of the adequately-sized ordinary chondrites in the Vatican collection. They established that the techniques were not only reliable, non-destructive, and non-contaminating, but also fast—several meteorites could be measured in one ordinary work day.

Not long before, Pierre Rochette and Jérôme Gattacceca, two researchers from the Centre Européen de Recherche et d'Enseignement des Géosciences de l'Environnement (CEREGE) in France, had been invited to visit the Vatican Observatory to measure a different physical property of the meteorites in the collection: magnetic susceptibility. In these measurements, the meteorite is placed inside electrical coils that produce an oscillating magnetic field. Paramagnetic and other magnetic materials within the meteorite respond to this by producing their own magnetic field. This in turn affects the frequency of the oscillation, which is measured. Magnetic susceptibility serves as a good indicator of the amount of metal and other magnetic materials are present within the meteorite. The three major classes of ordinary chondrite, H (high iron), L (low iron), and LL (low iron, low metal), were known to occupy different magnetic susceptibility ranges, though with some overlap (Rochette et al. 2003).

Since magnetic susceptibility data existed for many of the stones with the new density and porosity data, Bros. Consolmagno and Macke observed that, in a plot that contained both grain densities and magnetic susceptibilities for ordinary chondrite falls, the three groups occupied distinct regions of the graph (Fig. 10). They demonstrated that grain density in conjunction with magnetic susceptibility could be used as a reliable first-order classification technique for ordinary chondrite falls, without the need for invasive, destructive analyses. (For finds, however, terrestrial weathering of iron metal produced oxidized minerals that were both lower in density and lower in magnetic susceptibility, skewing the graph and rendering the technique inapplicable.) (Consolmagno et al. 2006)

Later, Br. Macke began a graduate program at University of Central Florida where for his doctoral dissertation he applied this same technique to over 1200 meteorite specimens from several institutional collections around the world, making it the largest survey to date of meteorite density and porosity. (cf. Macke 2010; Macke et al. 2010; Macke et al. 2011a, b) He also constructed a new ideal-gas pycnometer for specimens up to 10 cm diameter × 12 cm, with inserts for thin slabs up to 0.5 cm thick (cf. Macke et al. 2013; Fig. 8c), which now resides in the meteorite laboratory. In 2013, Br. Macke formally joined the staff of the Vatican Observatory, and, with the appointment of Br. Consolmagno to become president of the Vatican Observatory Foundation, Br. Macke took up the mantle of curator in 2014.

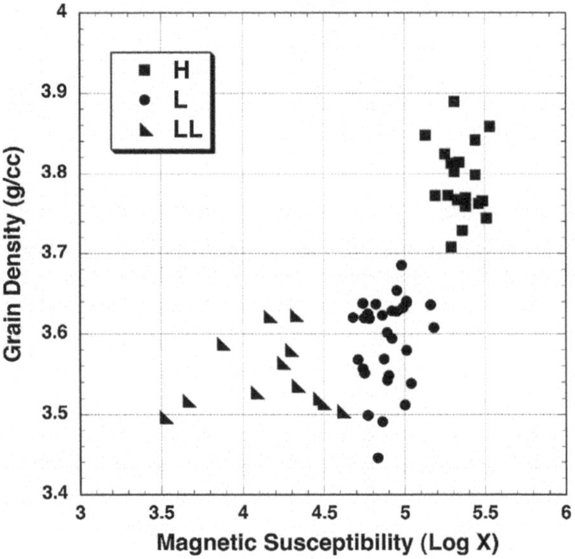


**Fig. 10** Grain density versus magnetic susceptibility for ordinary chondrite falls, with magnetic susceptibility presented in log SI units. The three populations of H, L, and LL ordinary chondrite falls separate into distinct regions (*Source* Consolmagno et al. 1998)

In the fall of 2014, Br. Macke encountered some issues with the glass bead methods that prompted a shift to other techniques. Part of his research involved the density and porosity of lunar materials, including specimens recovered during the Apollo missions. The Apollo specimen allocation body, CAPTEM, raised concerns about the sodium content in the glass beads and required alternative methods be applied for measuring bulk density. It happened that there was a 3-D laser scanner on-site at NASA Johnson Space Center that proved adequate for the task.

In 3-D laser scanning, volume is calculated from a computer model of the specimen. The use of laser scanning for meteorites and related materials has been a long time coming (for instance, see early work by Herd et al. 2003, and McCausland et al. 2007), but its widespread use has had to wait for the development of better scanning technologies that could handle the low-albedo (dark) surfaces of fusion crusts, sharp edges of cut or broken faces, and in some cases specular surfaces as well. By 2014, affordable laser scanner technology had improved to the point where these problems were surmountable. This led in 2015 to the acquisition by the Vatican Observatory of a NextEngine ScannerHD laser scanner (Fig. 11) through collaboration with Dan Britt. This particular device produces scans of better quality, sensitivity, and resolution than the low-end scanners, but is at least an order of magnitude less expensive than the high-end devices.

Laser scanning, while much slower than the bead method and requiring much more post-scan processing of data, offers several advantages. First, it reduces contact with the sample, allowing the scanning of very friable samples. One of the first pieces placed under the laser scanner was the large piece of the CI carbonaceous chondrite Orgueil, which is so friable that it is generally not handled at all. (The scanner completed the scan without damaging the specimen.) Second, it can potentially be as much as an order of magnitude more precise than the beads. This is particularly true when applied to dense samples with flat faces, which can compress beads in peculiar ways not well accounted for. Third, workbenches and other nearby surfaces tend to stay cleaner when lasers are used rather than glass beads. Due to static electricity, the beads had a way of migrating out of their confines and onto all surrounding surfaces.

The techniques developed by Britt and Consolmagno are today standard techniques for the measurement of meteorite density and porosity, applied widely by scientists at other institutions. They have the benefit of not only being applicable to meteorites, but to any solid material with well-interconnected interior pore space. (If the pores are not interconnected, then the gas used for ideal-gas pycnometry will not be able to penetrate all of them, thus skewing results. This is not a problem for most meteorites due to intergranular pores in matrix and widespread interconnected shock cracks.)

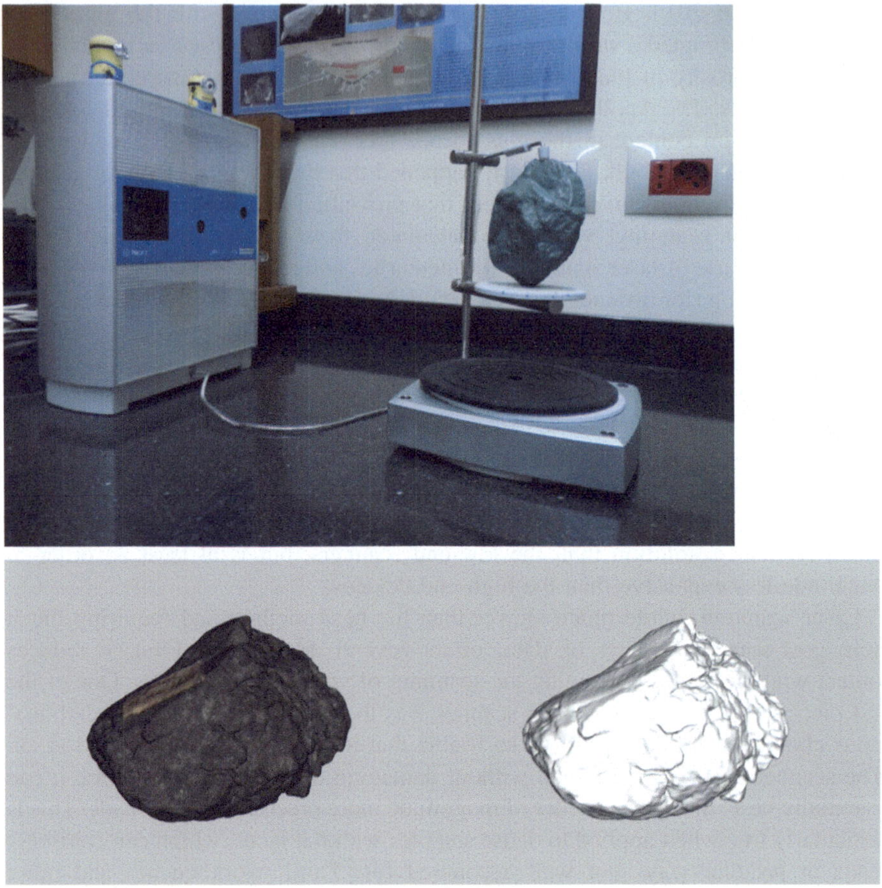

**Fig. 11** (top) NextEngine ScannerHD 3D laser scanner used at the Specola since 2015 to measure meteorite bulk volume and density. (below) 3D computer models of the Vatican specimen of the CI chondrite Orgueil produced from laser scans made by the NextEngine ScannerHD. This specimen is too friable to have been immersed in glass beads as had been used in earlier bulk density measurement techniques

## *Thermal Properties*

More recently, the focus of research on the Vatican meteorite collection has expanded to include thermal properties; in particular, heat capacity ($C_p$) and thermal conductivity ($k$). Both of these physical properties are integral in other thermal properties, such as thermal inertia $\Gamma = \sqrt{\rho C_p k}$ and thermal diffusivity $\kappa = k/(\rho C_p)$. When applied to asteroids, these properties influence effects that depend on how heat is stored and transported within the asteroid. For instance, the dynamics of asteroid orbits and spin is affected by Yarkovsky effects, in which sunlight is absorbed at subsolar surfaces, which are carried away from the subsolar point by the asteroid's rotation, and subsequently reradiate the energy in a different direction providing a small but significant change in momentum of the body. Thermal diffusivity is necessary for understanding the thermal evolution of the interior of the asteroid, which in turn dictates when and how certain minerals could form.

Until recently, there had been very little hard data on meteorite thermal properties, due in part to the difficulty in measuring these properties reliably. The measurement of thermal conductivity on Vatican meteorites began in 2010 in collaboration with Fr. Cyril Opeil S.J. at Boston College (Opeil et al. 2010). Fr. Opeil cut small samples of Vatican meteorites into parallelepipeds of about 0.3–1.0 cm per side and measured the thermal conductivity across the length of each structure. He used a Quantum Design Physical Property Measurement System, Thermal Transport Option (PPMS) (Fig. 12). The PPMS system with cryogenic cooling is capable of measuring thermal conductivity as a function of temperature over a range of 5–400 °C (cf. Fig. 13). To date, he has applied this measurement

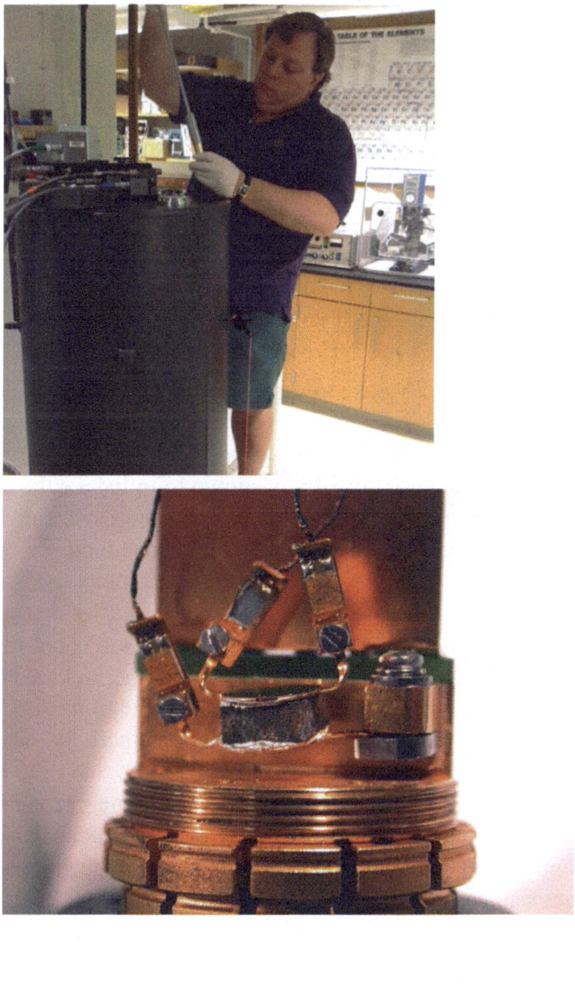

**Fig. 12** (above) Fr. Cyril Opeil S.J. with the quantum design physical properties measurement system—thermal transport option (PPMS) at Boston College. (below) A small piece of a meteorite in the PPMS sample mount (*Source* C. Opeil S.J.)

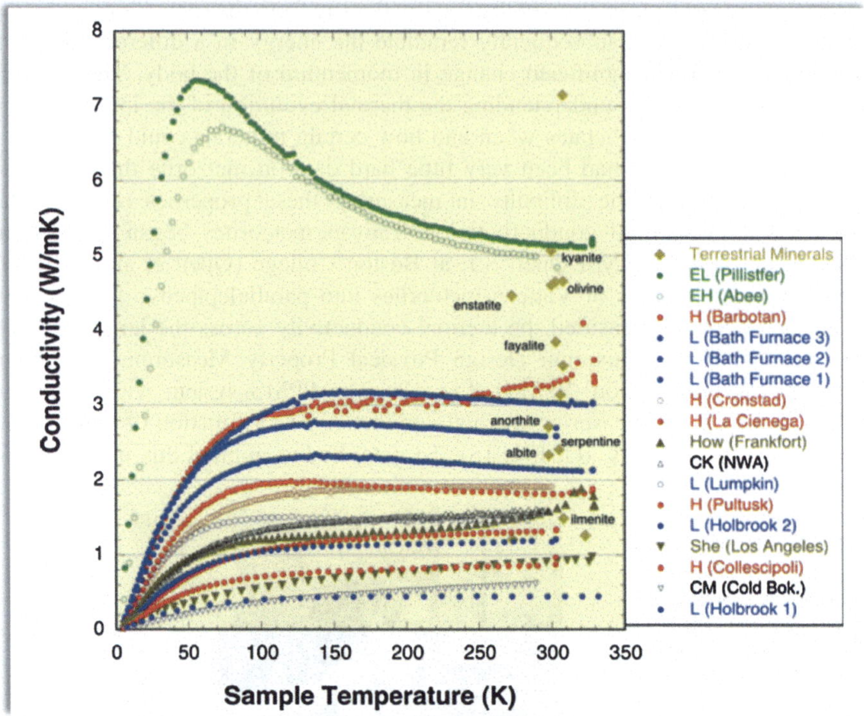


**Fig. 13** Some of the meteorite thermal conductivity data as a function of temperature collected by the PPMS system (taken from Opeil et al. 2012). Since this graph was first published, data for several more meteorites have been collected

technique on 26 meteorite samples, the majority of which were from the Vatican collection.

Thermal conductivity is a property that is not merely influenced by composition, but also structure. Porosity is known to influence the efficiency with which heat will travel across the sample. (Yomogida and Matsui 1983; Opeil et al. 2012). In addition, simple first-order thermal conductivity simulations (cf. Macke and Consolmagno 2014) show that pore geometry is also an important factor, particularly when the pore space is characterized by oriented shock cracks, as might be the case when the parent body has been subjected to significant impacts. Thermal conductivity along the direction traversing cracks will be significantly less than the thermal conductivity along directions parallel to the cracks. In the laboratory work, Fr. Opeil measured different thermal conductivity curves for the same sample under different orientations (Opeil et al. 2012).

Heat capacity is another important thermal property for which very little useful data had existed. In the research of the Vatican Observatory, a few tiny (several $mm^3$) meteorite specimens have measured heat capacity as a function of temperature using the PPMS system at Boston College. To supplement those time

**Fig. 14** The apparatus for measuring heat capacity by LN2 immersion

consuming and expensive measurements, Br. Consolmagno developed a novel technique for measuring the average heat capacity over the range 77–290 K of whole-rock specimens by dropping them into liquid nitrogen (LN2). (Consolmagno et al. 2013; Fig. 14) In this technique, a straight-walled dewar of liquid nitrogen is placed on a scale which reports the mass of the apparatus every few seconds to a computer, which records the mass curve. The nitrogen will evaporate at a fairly steady state, determined from the evaporation curve in the minutes prior to beginning the measurement. A meteorite is then placed into the dewar. As its temperature drops from room temperature to the temperature of liquid nitrogen (77 K), the heat causes LN2 to rapidly boil away. After the temperature has settled at 77 K, the LN2 returns to a steady evaporation curve. By comparing the evaporation curve before and after the introduction of the sample, one can calculate the amount of LN2 that evaporated due to heat exchange with the meteorite, and using

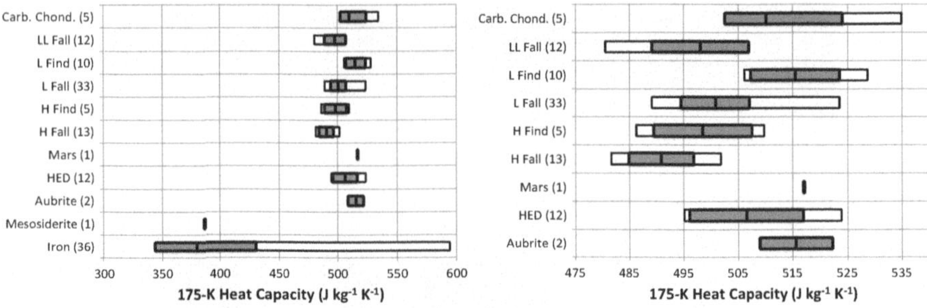


**Fig. 15** Summary of LN2-based heat capacity results. Shaded boxes are one-sigma from the mean (central bar), and open boxes cover the range of results for the population. Numbers in parentheses indicate the number of meteorites measured per group. Plot to the right is the same as to the left, but with irons and mesosiderites removed and the graph rescaled for clarity

the latent heat of vaporization of LN2 one can then determine the heat capacity of the sample. This technique yields a fairly good approximation of the heat capacity of the sample at 175 K.

To date, 51 meteorites have had Cp(T) measured by Fr. Opeil using the PPMS system, and more than 130 meteorites of the Vatican collection have been exposed to the LN2 immersion procedure by Bros. Consolmagno and Macke (Fig. 15). The meteorites submerged in LN2 have included iron meteorites, ordinary chondrites, and HEDs (thought to have originated on the asteroid Vesta) and several others.

## Summary

In addition to having unique historical value due to its origins in the French aristocracy and containing a number of historically relevant and interesting specimens, the Vatican meteorite collection continues to serve as a valuable scientific collection of meteorites. Through the work of the Jesuits of the Observatory and their collaborators, the Vatican meteorite collection provides a valuable source of material not only for existing research programs (including loans of specimens and thin sections provided to researchers at other institutions doing more traditional meteoritics research), but also as a developmental test bed for new measurement techniques. The Vatican meteorite collection played a significant role in the early work in spectroscopy on extraterrestrial materials, in the development of non-destructive and non-contaminating techniques for measuring meteorite density and porosity, and in the development of novel new ways to measure low-temperature thermal properties on whole stones. In doing so, it filled a particular scientific niche in developing lines of research that were not being pursued at other institutions despite the clear need for the data these measurements provide. The collection continues to grow slowly, and with luck and Providence will continue to serve as a valuable source of research material for the development of new techniques in the years to come.

**Acknowledgements** I would like to thank Br. Guy Consolmagno S.J. and Fr. Sabino Maffeo S. J. for their earlier work in tracing the history of the collection before the 1980s. Much of the first part of this article benefited greatly from their work: Consolmagno (2006) and Maffeo (2001).

## References

Brant, S. (1492). *Von dem donnerstein gefallen im 1492 jar: Vor Ensisheim* (1 p). Strasbourg: Johannes Bruss.

Cadenhead, D. A., & Stetter, J. R. (1975). Specific gravities of lunar materials using helium pycnometry. In *Proceedings of the 6th Lunar Science Conference* (pp. 3199–3206). New York: Pergamon Press, Inc.

Cadenhead, D. A., Wagner, N. J., Jones, B. R., & Stetter, J. R. (1972). Some surface characteristics and gas interactions of Apollo 14 fines and rock fragments. In *Proceedings of the Third Lunar Science Conference* (pp. 2243–2257). Cambridge, MA: M.I.T. Press.

Consolmagno, G. (2006). A brief history of the Vatican meteorite collection. *Geological Society, London, Special Publications, 256,* 205–217.

Consolmagno, G. J., & Britt, D. T. (1998). The density and porosity of meteorites from the Vatican collection. *Meteoritics & Planetary Science, 33,* 1231–1241.

Consolmagno, G. J., Britt, D. T., & Stoll, C. P. (1998). The porosities of ordinary chondrites: Models and interpretation. *Meteoritics & Planetary Science, 33,* 1221–1229.

Consolmagno, G. J., Macke, R. J., Rochette, P., Britt, D. T., & Gattacceca, J. (2006). Density, magnetic susceptibility, and the characterization of ordinary chondrite falls and showers. *Meteoritics & Planetary Science, 41,* 331–342.

Consolmagno, G. J., Schaefer, M. W., Schaefer, B. E., Britt, D. T., Macke, R. J., Nolan, M. C., et al. (2013). The measurement of meteorite heat capacity at low temperatures using liquid nitrogen vaporization. *Planetary and Space Science, 87,* 146–156.

Fujii, N., Miyamoto, M., Kobayashi, Y., & Ito, K. (1981). Differences of relative strength among chondrites measured by the vibrational fracturing rate. *Memoirs of the National Instituite of Polar Research, 20,* 362–371.

Gatterer, A., & Junkes, J. (1940). Über den Steinmeteoriten von Rio Negro. *Commentationes Pontificia Academia Scientiarum, 4,* 191–223.

Geddis, A. M. (1994). *Rapid estimate of solid volume in large tuff cores using a gas pycnometer* (M.S. thesis), 118 pp. University of Arizona.

Herd, R., Spray, J., Samson, C., Miller, S., & Christie, I. (2003). 3D imaging and modelling with a space-qualified laser camera system: Development of terrestrial applications and potential for planetary exploration (abstract #1718). In *34th Lunar and Planetary Science Conference.*

Ireland, T. R., Compston, W., Williams, I. S., & Wendt, I. (1990). U-Th-Pb systematics of individual perovskite grains from the Allende and Murchison carbonaceous chondrites. *Earth and Planetary Science Letters, 101,* 379–387.

Keil, K. (1962). Quantitativ-erzmikroskopische Integrationsanalyse der Chondrite. *Chemie der Erde, 22,* 281–348.

Macke, R. J. (2010) *Survey of meteorite physical properties: density, porosity and magnetic susceptibility* (313 pp). Ph.D. thesis. Orlando, FL: University of Central Florida.

Macke, R. J., Britt, D. T., & Consolmagno, G. J. (2011a). Density, porosity and magnetic susceptibility of achondritic meteorites. *Meteoritics & Planetary Science, 46,* 311–326.

Macke, R. J., Consolmagno, G. J., & Britt, D. T. (2011b). Density, porosity and magnetic susceptibility of carbonaceous chondrites. *Meteoritics & Planetary Science, 46,* 1842–1862.

Macke, R. J., Britt, D. T., & Consolmagno, G. J. (2013). New pycnometer design for thin-sliced meteorites (abstract #1398). In *44th Lunar and Planetary Science Conference.*

Macke, R. J., & Consolmagno, G. J. (2014). Porosity and pore geometry influence on meteorite thermal conductivities (abstract #5045). In *77th Annual Meteoritical Society Meeting.*

Macke, R. J., Consolmagno, G. J., Britt, D. T., & Hutson, M. L. (2010). Enstatite chondrite density, magnetic susceptibility and porosity. *Meteoritics & Planetary Science, 45,* 1513–1546.

Maffeo, S. (2001). *La Specola Vaticana: Nove papi, una missione* (p. 416). Vatican City: Libreria Editrice Vaticana.

Marvin, U. B., Wood, J. A., Dickey, J. S. Jr. (1970). Ca-Al rich phases in the Allende meteorite. *Earth and Planetary Science Letters, 7,* 346–350.

Matsui, T., Hamano, Y., & Honda, M. (1980). Porosity and compressional-wave velocity measurement of Antarctic meteorites. *Memoirs of the National Institute of Polar Resarch, 17,* 268–275. (special issue).

McCausland, P. J., Samson, C., & DesLauriers, A. (2007). Bulk densities of meteorites via visible light 3D laser imaging. *Meteoritics & Planetary Science Supplement, 42,* 5066.

Opeil, C. P., Consolmagno, G. J., & Britt, D. T. (2010). The thermal conductivity of meteorites: New measurements and analysis. *Icarus, 208,* 449–454.

Opeil, C. P., Consolmagno, G. J., Safarik, D. J., & Britt, D. T. (2012). Stony meteorite thermal properties and their relationship with meteorite chemical and physical states. *Meteoritics & Planetary Science, 47,* 319–329.

Pesonen, L., Terho, M., & Kukkonnen, I. T. (1993) Physical properties of 368 meteorites: Implications for meteorite magnetism and planetary geophysics. In *Proceedings of the NIPR Symposium of Antarctic Meteorites* (vol. 6, pp. 401–416). Tokyo: National Institute of Polar Research.

Rochette, P., Sagnotti, L., Consolmagno, G., Denise, M., Folco, L., Gattacceca, J., Osete, M., Pesonen L. (2003) Magnetic classification of meteorites (abstract #5975). In *EGS-AGU-EUG Joint Assembly.*

Ruzicka, A., Grossman, J., Bouvier, A., Herd, C. D. K., & Agee, C. B. (2015). *The Meteoritical Bulletin, 102,* 9.

Salpeter, E. (1952). Spektroskopische Chlorbestimmung in Steinmeteoriten. *Richerche Spettroscopiche, 2,* 1–63.

Stein, G. (1933). Notizie sul trasferimento e sulla riorganizzazione della Specola Vaticana. *Atti della Pontificia Accademia delle Scienze Nuovi Lincei, 86,* 24–32.

Tizio, S. (1528) *Historia Senensium (History of the sienese)* Biblioteca Apostolica Vaticana, MS Chigi G.II.36.

Yomogida, K., & Matsui, T. (1981). Physical properties of some Antarctic meteorites. *Memoirs of the National Institute of Polar Research, 20,* 284–294.

Yomogida, K., & Matsui, T. (1982). Physical properties of some unequilibrated Antarctic ordinary chondrites. *Memoirs of the National Institute of Polar Research, 25,* 308–318. (special issue).

Yomogida, K., & Matsui, T. (1983). Physical properties of ordinary chondrites. *Journal of Geophysical Research, 88,* 9513–9533.

# Near Earth Objects and Their Physical Characterization

**Jean-Baptiste Kikwaya Eluo, S.J.**

**Abstract** The present research on Near Earth Objects (NEOs) is mostly limited on addressing the question on their number and how hazardous they may be to the Earth. Catalina Sky Survey (CSS) has submitted thousands of astrometric observations to the Minor Planet Center (MPC) resulting in discovery of hundreds of NEOS but much of their physical characteristics remain unknown. OSIRIS-Rex, a return sample mission, will visit 1999 RQ36, a carbonaceous asteroid, physical characteristic which could only be identified through other techniques of observations than just astrometry. Using VATT (Vatican Advanced Technology Telescope), we observed 11 NEOs using photometry with four broadband filters (BVRI). With this project where we combined two approaches (two-color plots, and comparison of relative reflectance normalized to V-filter with the real observed spectra), we were able to determine color of the 11 NEOs, and therefore had some clue about their physical characteristics particularly composition, size, structure, albedo…

## Introduction

Until a few years ago, research on Near Earth Objects (NEOs) was solely oriented toward addressing the question of their number and how hazardous they might be to the Earth. For example, since 1998 the Catalina Sky Survey (CSS) has been searching for NEOs using 68/76 cm f/1.9 classical Schmidt optics capable of going down to 20.5 magnitude, and it has submitted thousands of astrometric observations to the Minor Planet Center (MPC) resulting in discovery of hundreds of NEOs. But much of their physical characteristics remain unknown. On the other hand, OSIRIS-Rex, a return sample mission launched on October 8th, 2016, will visit 1999 RQ36, an NEO asteroid (asteroid Bennu) which is thought to be a carbona-

J.-B. Kikwaya Eluo, S.J. (✉)
Vatican Observatory, Tucson, AZ 85721, USA
e-mail: jbkikwaya@gmail.com



G. Gionti, S.J. and J.-B. Kikwaya Eluo, S.J. (eds.), *The Vatican Observatory, Castel Gandolfo: 80th Anniversary Celebration*, Astrophysics and Space Science Proceedings 51, https://doi.org/10.1007/978-3-319-67205-2_2

ceous asteroid: a physical characteristic that could only be identified through other techniques of observation than just astrometry.

Asteroids are the building blocks of terrestrial bodies (Mercury, Venus, Earth, and Mars), and possibly also of the cores of the giant planets (Jupiter, Saturn, Uranus, and Neptune). They conserve the signature of the presolar nebula in which the planets formed. They can also inform us on how life started on Earth since they might contain the original organic matter and water. In fact, we know now, thanks to Rosetta mission, comets might probably not be the source of Earth's water. There is then a real need to focus on asteroids, which must be studied in order to answer questions related to the origin of water and life on Earth. And that can be accomplished when we begin to study the physical characterization of asteroids.

As asteroids carry the signature of the birth of the solar system, they are able to help us also to understand how many stars injected their matter into early solar system and what kind of stars they might be. As we gather observed compositional and structural properties of asteroids, we enable ourselves more and more to build new theories to explain the different processes that brought the solar system from its initial state to the present one. This knowledge can also allow us to infer what is happening in other solar systems at different stages of their evolution.

The most accurate method to study the physical characterization of asteroids is to use spectroscopy that covers both the optical and near-infrared electromagnetic spectrum to obtain their reflectance spectra, and thermal infrared emission to estimate their albedo (Dandy et al. 2002). This method not only is time-consuming, but it is also very expensive.

A very straight forward and inexpensive method is to use optical photometry with a set of very broad band filters from Blue to Infrared (BVRI) (Landolt 1992). The four main asteroid spectral classes from the Tholen classification (C, D, M, and S) are easily obtained with this method (Zellner et al. 1985). C represents dark carbonaceous asteroids. D groups all those asteroids with low albedo. They contain organic-rich silicates, carbon, and possibly water ice. They are found mostly in the outer asteroid belt and beyond. Class M is a class of metallic asteroids. All "stony" asteroids are put together in class S.

In 2014 we started a program of observing NEOs using the Vatican telescope at Mt Graham in Arizona in USA. Our project consists in determining the physical characterization of NEOs. We use photometry with 4 broadband Johnson-Cousin filters. To determine the surface color of a particular NEO, we use the two color plots of three different color indices (B-V, V-R, and V-I) (Zellner et al. 1985). This method, as a first approach, helps to estimate the optical color of asteroids in general, and NEOs in particular (Dandy et al. 2003; Yoshida et al. 2004). In our work, we take this method one step further by creating from these color indices a very broad spectrum normalized at V color for each NEO, which we compare with real observed spectra in our database through the program "modeling for asteroids" by Birlan Mirel (Popescu et al. 2012). In this way, more information becomes available to describe physically each NEO than what the two-color plots alone can provide from optical photometry.

## Instrument and Data

We use Vatican Advanced Technology Telescope (VATT) for gathering our data. VATT is an f/1.0 telescope with a primary mirror of 1.8 m in diameter. Its secondary mirror is f/0.9 with a diameter of 0.38 m. For photometry and astrometry, we use a new STA0500A back illuminated 4K CCD camera with a resolution of 4064 × 4064 pixels on 15 × 15 μm. Without binning, the pixel scale is 0.188 arcsec/pixel and the image reading time is 60 s. For our project on NEOs, we set the CDD to bin two by two in order to reduce the readout time to 30 s. Consequently, the resolution dropped to 2016 × 2016 pixels and the pixel scale to 0.375 arcsec/pixel. The wavelength range covers the entire visible spectrum (300–1000 nm). The quantum efficiency of the instrument (VATT 4K) peaks at 450 nm.

Our project consists in observing those NEOs thought to have a fast spin rate (fast rotators). To ensure that we observe fast rotators, we aim at those asteroids with high absolute magnitude (H > 22) as they tend to be small, on the order of several dozen meters (Fedorets et al. 2017; Tricarico 2017; Schunova-Lilly et al. 2017). We collected 11 NEOs from 2014 to 2016 with H ranging from 21.5 to 27. They are 2014 AY28, 2014 EC, 2014 KS40, 2014 WF201 observed with only three filters (B, V, and R), 2015 FP with only two filters (V, and R), 2015 TB25, 2015 VM64, 2015 VT64, 2015 XZ1, 2016 EW1, and 2016 GW221 observed using four different filters (B, V, R, and I).

We will describe in great detail NEO 2014 WF201 (Fig. 1), and for others, we will only give the final results as we use the same technique to observe, reduce, and analyze them. 2014 WF201 was announced by the Minor Planet Center on November 24th, 2016. Its absolute magnitude was 25.6. Its delta distance (its closest distance to the Earth) was 0.018 AU (astronomical distance, distance between the Sun and the earth being 1 AU) when being at 1.005 AU from the Sun. We observed it the night of November 25th, 2016 and November 26th, 2016. We took 85 frames with the R-filter to cover its complete lightcurve in order to estimate its spin rate. For color, we observed 2014 WF201 with three more filters. We took 5 frames in the B-filter, 5 frames in the V-filter, and 5 frames the I-filter with the following sequence: R-B-R-V-R-I-R: 15 frames in R, 5 frames in B, 15 frames in R, 5 frames in V, 15 frames in R, 5 frames in I, and 40 frames in R. The details for the other 10 NEOs are reported in Table 1.

## Reduction and Analysis

For each campaign run, we took nearly 200 bias frames, and using the IRAF packages imred, ccdred, and zerocombine (Tody 1986) we created a master bias frame to be used in the calibration of the science frames. We also took 50 flat frames in each filter. One master flat in each filter was then created using the IRAF packages imred, ccdred, and flatcombine (Tody 1993). For the calibration of the

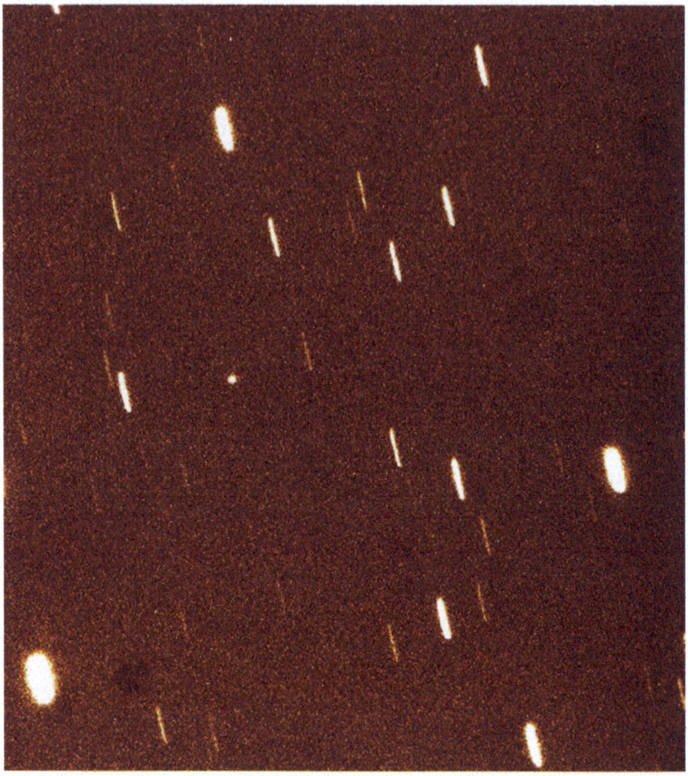

**Fig. 1** NEO 2015 WF201 appears as a dot. Stars are trailed because the image was taken following the bias rate of the object (motion of the object on the sky)

**Table 1** 11 NEOs, dates of observation, number of frames for each filter (R, B, I, and V), object absolute magnitude (H), distance from the Earth, and distance from the Sun as released by Minor Planet Center

| Asteroid | Date of observation | R-filter | B-filter | I-filter | V-filter | H | Delta (AU) | r (AU) |
|---|---|---|---|---|---|---|---|---|
| 2014AY28 | 2014 04 05 | 42 | 5 | – | 7 | 21.8 | 0.071 | 1.024 |
| 2014EC | 2014 03 05 | 40 | 50 | – | 50 | 28.2 | 0.017 | 1.009 |
| 2014KS40 | 2014 06 01 | 50 | 15 | – | 10 | 21.9 | 0.068 | 1.065 |
| 2014WF201 | 2014 11 26 | 85 | 5 | 5 | 5 | 25.6 | 0.018 | 1.005 |
| 2015FP | 2015 03 24 | 75 | – | – | 15 | 25.2 | 0.031 | 1.027 |
| 2015TB25 | 2015 10 14 | 135 | 10 | 10 | 10 | 24.5 | 0.027 | 1.019 |
| 2015VM64 | 2015 11 09 | 50 | 4 | 4 | 4 | 26.1 | 0.025 | 1.015 |
| 2015VT64 | 2015 11 09 | 60 | 5 | 6 | 5 | 26.0 | 0.025 | 1.015 |
| 2015XZ1 | 2015 12 08 | 20 | 5 | 5 | 5 | 25.0 | 0.043 | 1.028 |
| 2016EW1 | 2016 03 07 | 45 | 4 | 4 | 4 | 25.2 | 0.028 | 1.017 |
| 2016GW221 | 2016 05 03 | 45 | 5 | 5 | 5 | 24.8 | 0.048 | 1.049 |

science frames observed with one particular filter, we applied master bias and master flat of the same filter to them using the IRAF packages imred, and ccdproc.

For the photometry reduction, we used the IRAF packages ccdphot, digiphot, apphot. For one particular frame, we detected the dot representing the object, we used the IRAF command imexam combined with the r-key to get the object profile that gives among other characteristics of the object its full width at half maximum (FWHM). We set the aperture for the photometry to twice this value with the IRAF command photpars.aperture = (2 * FWHM). The IRAF command phot followed by the name of the frame fits file gives, the time that frame was captured in UT, the magnitude of the object on that particular frame, the error in magnitude, the exposure time, the airmass, and the pixel value of the sky.

We also observed standard stars to correct for the fact that we can't use the differential photometry, as the stars in the field don't appear as point sources due to our observation settings. We take 4–5 standard stars observed with each of the four filters. After applying bias and flat calibration, we calculate total magnitude of each standard star as:

$$\begin{aligned}\text{Total magnitude} &= \text{Standard star catalog magnitude}\\ &\quad - \text{Standard star Instrumental magnitude}\end{aligned}$$

We plot total magnitude of the standard stars versus their air masses and make a linear fit whose slope is the extinction coefficient. Typical extinction coefficients for VATT4K are 15% for B-filter, 10% for V-filter, 8% for R-filter, and 5% for I-filter. To apply the standard stars correction to the science frames, we use the zero point and the slope from the linear fit. The magnitude of the object on each frame is modified as:

$$\begin{aligned}\text{Object magnitude on one frame} &= (\text{zero point magnitude} + \text{object instrumental magnitude})\\ &\quad - \text{object airmass} \times \text{extinction coefficient}\end{aligned}$$

We obtain as many magnitude measurements of the object as the number of frames in each of the four filters. We used ALC (Asteroid Light Curve), an Asteroid Lightcurve Analysis Program, to compute both the lightcurve and the color indices of the object (Pravec, x). We begin by displaying the frames taken with the R-filter. When the B-filter frames are displayed, there is an offset with the R-filter, which is the color index R-B. We adjust the B-filter object measurements to the R-filter ones. And we proceed with the V-filter, and I-filter to have the color indices R-V, and R-I. We transform these color indices to B-V, V-R, and V-I as the two color plots are B-V versus V-R, and V-I versus V-R (Table 2). The error bar in each filter is generated from the lightcurve fit to data.

For each NEO, we computed the absolute magnitude using its orbital elements (semi-major axis, eccentricity, inclination, perihelion, and period) (Table 3). We placed the 11 NEOs in the a–e plot as suggested by Dandy et al. 2003 where Aten, Apollo, Amor, and Jupiter family comets ($2 < T < 3$) defined by Bottke et al.

**Table 2** 11 NEOs with their color indices V-R, B-V, and V-I that help to make the two color plots in order to estimate in a very broad and first approach the surface color of the asteroids: type C, type D, type M, and type S (Zellner et al. 1985)

| Asteroid | V-R | d(V-R) | B-V | d(B-V) | V-I | d(V-I) |
|---|---|---|---|---|---|---|
| 2014AY28 | 0.32 | 0.045 | 0.793 | 0.051 | – | – |
| 2014EC | −0.052 | 0.099 | 0.567 | 0.087 | – | – |
| 2014KS40 | 0.445 | 0.053 | 0.595 | 0.052 | – | – |
| 2014WF201 | 0.399 | 0.036 | 0.795 | 0.062 | 0.688 | 0.041 |
| 2015FP | 0.258 | 0.051 | – | – | – | – |
| 2015TB25 | 0.338 | 0.052 | 0.805 | 0.052 | 0.694 | 0.054 |
| 2015VM64 | 0.315 | 0.047 | 0.768 | 0.044 | 0.751 | 0.046 |
| 2015VT64 | 0.333 | 0.044 | 0.895 | 0.054 | 0.626 | 0.053 |
| 2015XZ1 | 0.365 | 0.055 | 0.732 | 0.041 | 0.695 | 0.051 |
| 2016EW1 | 0.42 | 0.041 | 0.733 | 0.051 | 0.62 | 0.041 |
| 2016GW221 | 0.343 | 0.048 | 0.775 | 0.042 | 0.708 | 0.042 |

Two NEOs, 2014 EC and 2015 FP, are particularly of great interest as their color indices fall outside the ranges

**Table 3** Orbital elements of the 11 NEOs as given by Minor Planet Center

| Asteroid | Semi-major axis (AU) | Eccentricity | Inclination (°) | Perihelion (AU) | Period (year) |
|---|---|---|---|---|---|
| 2014AY28 | 1.425 | 0.283 | 5.709 | 1.023 | 1.70 |
| 2014EC | 1.459 | 0.526 | 1.403 | 0.691 | 1.76 |
| 2014KS40 | 2.294 | 0.538 | 2.924 | 1.060 | 3.47 |
| 2014WF201 | 1.109 | 0.109 | 8.505 | 0.988 | 1.17 |
| 2015FP | 1.550 | 0.348 | 9.712 | 1.010 | 1.93 |
| 2015TB25 | 1.282 | 0.208 | 16.915 | 1.016 | 1.45 |
| 2015VM64 | 1.886 | 0.619 | 16.311 | 0.718 | 2.59 |
| 2015VT64 | 1.166 | 0.171 | 17.988 | 0.967 | 1.26 |
| 2015XZ1 | 1.089 | 0.179 | 19.827 | 0.894 | 1.14 |
| 2016EW1 | 1.038 | 0.268 | 18.898 | 0.760 | 1.06 |
| 2016GW221 | 0.827 | 0.268 | 3.655 | 0.605 | 0.75 |

(2002) are well marked. Different taxonomy types tend to occupy different regions of (a, e) space (Dandy et al. 2003). The (a, e) plot also gives an indication of the source of NEOs.

NEOs belonging to Amor region with $a > 2$ or near 2.5 are believed to be delivered to the Earth trough the 3:1 resonance with Jupiter (Bottke et al. 2002) while those from Apollo and Aten regions reach the Earth through the v6 secular resonance. It happens that the 3:1 and v6 resonances are very "fast track" in the delivery of NEOs to the Earth (Bottke et al. 2002). Only $\sim 10^6$ years are needed for an object, entering this region as result of collision, to be delivered to the Earth (Froeschle et al. 1995; Rabinowitz 1998). We expect then that these NEOs would keep some freshness of their surface when they reach the Earth. And they tend to belong to D-type (also V-type, and R-type not described in this work). Their spectra show a deep 1-μm absorption band (Dandy et al. 2003).

NEOs from an IMC (Intermediate Mars-Crossing) source follow a very slow track of delivery (Rabinowitz 1997) as they take about $10^9$ years to reach the Earth. As result, they experience more the effect of space weathering. They tend then to be S-type (Dandy et al. 2003).

C-type NEOs seem to come from all over the place. Some place them among Amors with A > 1.5 AU and Apollos with moderate eccentricity (Dandy et al. 2003). We found them in the main belt from the central to the outer region where the delivery mechanism is dominated by the 3:1 resonance (Bottke et al. 2002). But they are also present in the region with a < 2 where we expect to find E or M-types (Dandy et al. 2003).

# Results and Discussion

## *Lightcurve and Spin Rate*

2014 WF201 lightcurve (Fig. 2) computed using ALC software determined its spin rate that is 1.00 ± 0.03 h. With this small rotation period, 2014 WF201 is considered to be a fast rotator. Error on each observed point of the lightcurve is estimated by combining error from the reduction and the one from the fit.

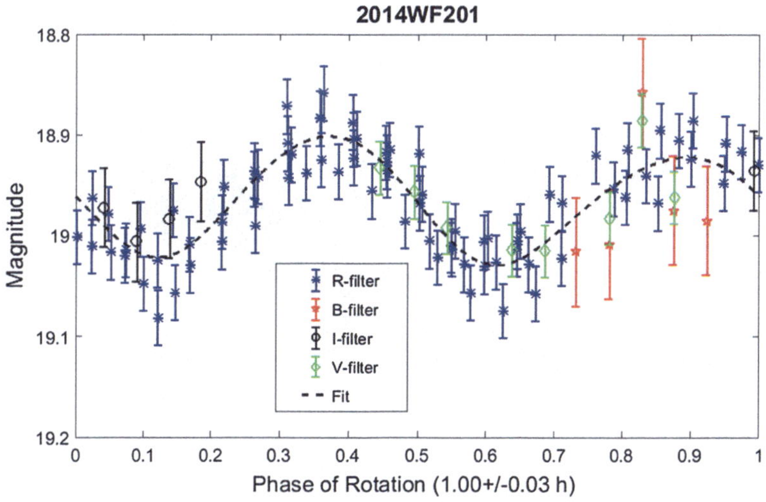


**Fig. 2** Lightcurve of neo 2014 WF201. The object came very close to the Earth at 0.018 AU

**Table 4** Spin rates of the 11 NEOs

| Asteroid | Rot. period (h) | Amplitude (mag) |
|---|---|---|
| 2014AY28 | 1.88 ± 0.03 | 0.32 ± 0.04 |
| 2014EC | 0.56 ± 0.01 | 0.48 ± 0.16 |
| 2014KS40 | 0.98 ± 0.04 | 1.26 ± 0.16 |
| 2014WF201 | 1.00 ± 0.03 | 0.13 ± 0.02 |
| 2015FP | 0.17 ± 0.01 | 0.51 ± 0.06 |
| 2015TB25 | 0.67 ± 0.04 | 0.31 ± 0.08 |
| 2015VM64 | 2.30 ± 0.03 | 0.23 ± 0.07 |
| 2015VT64 | 1.13 ± 0.01 | 0.40 ± 0.13 |
| 2015XZ1 | 0.95 ± 0.07 | 0.33 ± 0.09 |
| 2016EW1 | 1.34 ± 0.02 | 1.40 ± 0.03 |
| 2016GW221 | 0.29 ± 0.05 | 0.41 ± 0.10 |

For the 10 other objects, we give only their spin rates that range from 0.17 ± 0.01 h to 2.30 ± 0.03 h (Table 4) and the amplitudes of the lightcurves from 0.23 ± 0.07 mag to 1.40 ± 0.03 mag.

## *Two Color Plots*

In order to determine the color of any NEO in a very first approach, we used two color plots (V-R, B-V) and (V-R, V-I) from the work of Zeller et al. (1987) which gives the four main optical types of asteroids (C-type, D-type, M-type, and S-type). The two color plots of 2014 WF201 show that it belongs to the C-type. In fact,

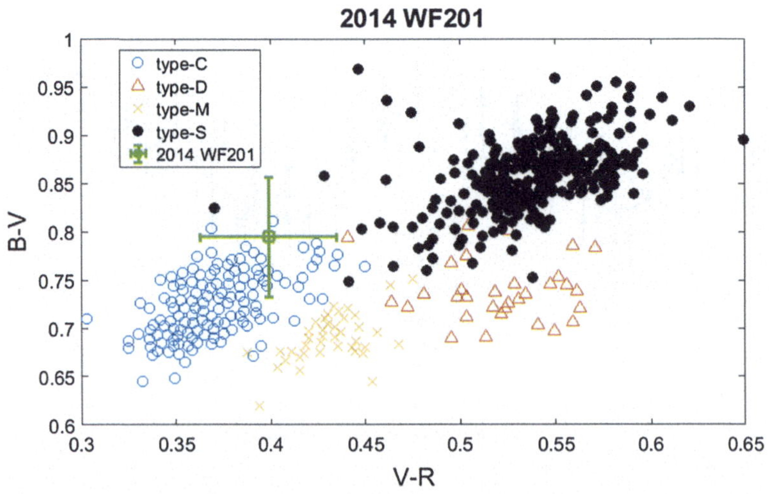


**Fig. 3** NEO 2014 WF201 is a C-type asteroid

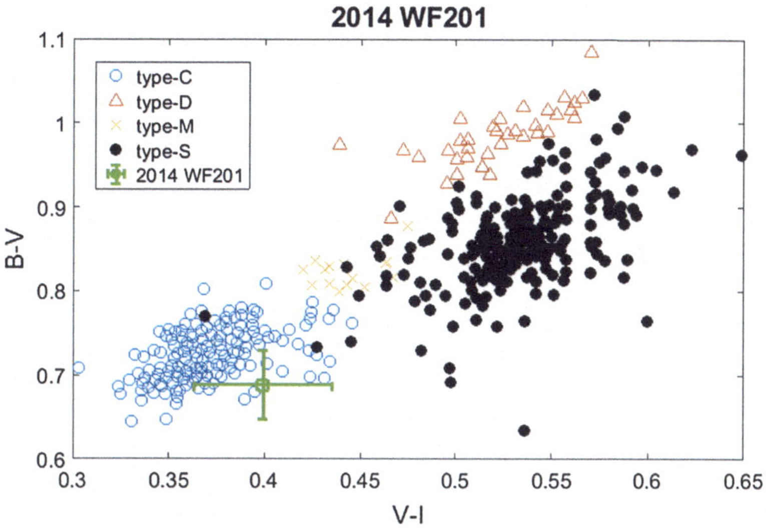


**Fig. 4** NEO 2014 WF201 falling among C-type asteroids is therefore a C-type object

in the two color plot B-V versus V-R, 2014 WF201 falls with C-type asteroids (Fig. 3), and this is confirmed by the two color plot V-I versus V-R where 2014 WF201 is also among C-type asteroids. For the 10 other objects, 6 are clearly C-type, one is either C-type or S-type, two from this method cannot be unequivocally classified, and one, being observed only with two colors (V, and R) cannot be placed in the two color plots (Fig. 4).

## *Relative Reflectance and Spectra*

From the color indices (V-R, B-V, and V-I), we computed reflectance values normalized to V of the 11 NEOs and we compared them with the observed spectra in database using the program "modeling for asteroids" by Birlan Mirel (Popescu et al. 2012). When comparing the optical relative reflectance of 2014 FW201 (Fig. 5) with spectra in the database, three different spectra match it (Figs. 6, 7 and 8). It turns out that 2014 WF201, which was optically described as C-type, can actually now be classified as Cg, Cgh (C-complex classes that are refined classes of carbonaceous asteroids) (Popescu et al. 2016). One representative of these classes is asteroid 1 Ceres. 2014 WF201 might also be an O-type, a very rare class of asteroids. The spectra of O-type asteroids match those of L6 and LL6 ordinary chondrite meteorites (Bus and Binzel 2002a, b).

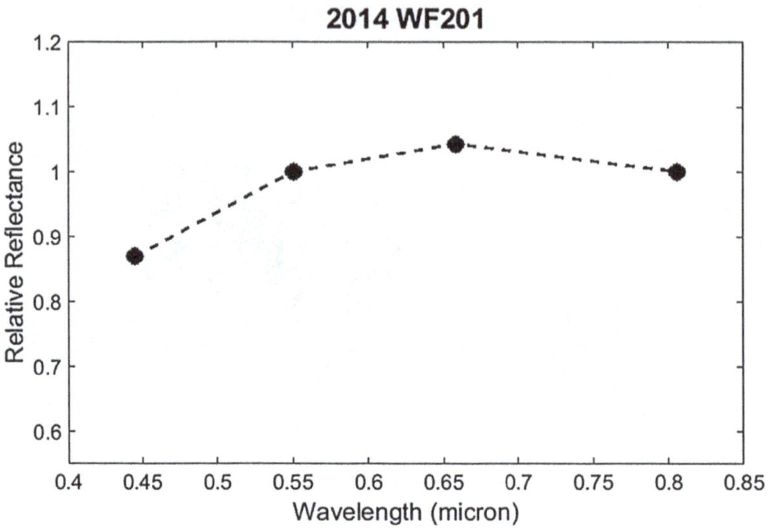


**Fig. 5** Optical reflectance of NEO 2014 WF201

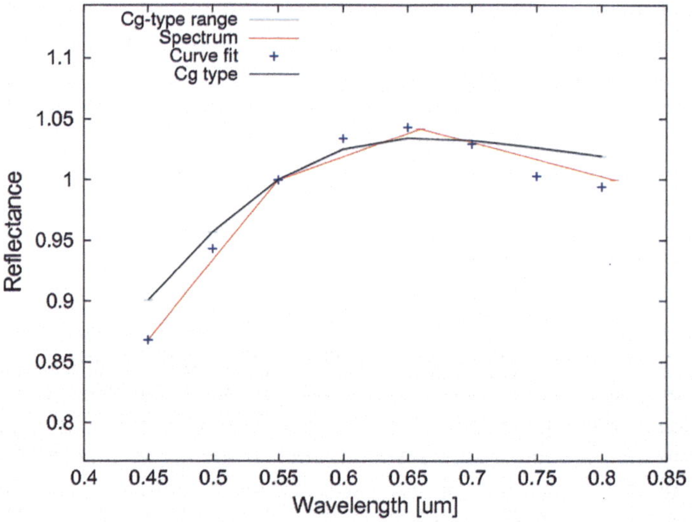


**Fig. 6** Optical reflectance of 2014 WF201 matches Cg-type spectra

Among the 11 NEOs, only three (2015 FP, 2014 KS40, 2014 EC) could not have their reflectance matching spectra in the database. Beside classes Cgh, Cg, and O, optical reflectance of 2015 VM64 matched spectra of Ch asteroids (Ch belongs also to C-complex classes) (Table 5).

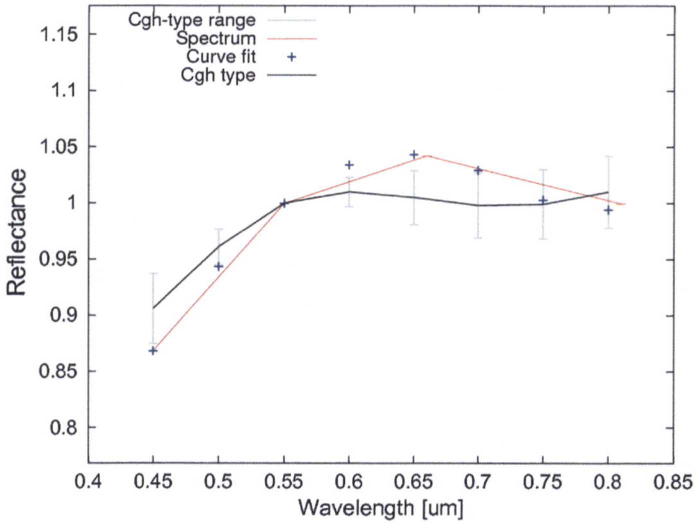


**Fig. 7** Optical reflectance of 2014 WF201 matches Cg-type spectra

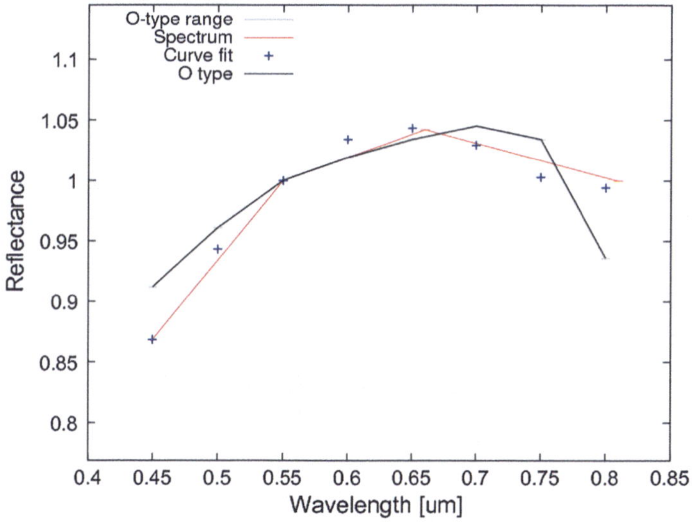


**Fig. 8** Optical reflectance of 2014 WF201 matches O-type spectra

## *Color and Orbits*

2014 AY28, 2014 WF201, 2015 TB25, 2015 VM64, 2015 VT64, 2015 XZ1, and 2016 GW221 are classified as C-type from the two color plots, and by comparing their relative reflectances with the observed spectra, they match various spectra (Cg,

**Table 5** The 11 NEOs classified both using the two color plots and the comparison with the observed spectra

| Asteroid | Classes | |
|---|---|---|
| | From Zeller et al. (1985) | From Popescu et al. (2012) |
| 2014AY28 | C | Cgh, Cg, O |
| 2014EC | N/A | – |
| 2014KS40 | X | – |
| 2014WF201 | C | Cg, Cgh, O |
| 2015FP | N/A | – |
| 2015TB25 | C | Cgh |
| 2015VM64 | C | Cgh, Ch |
| 2015VT64 | C | Cgh |
| 2015XZ1 | C | Cgh, C |
| 2016EW1 | C or S | O, Cg |
| 2016GW221 | C | Cgh |

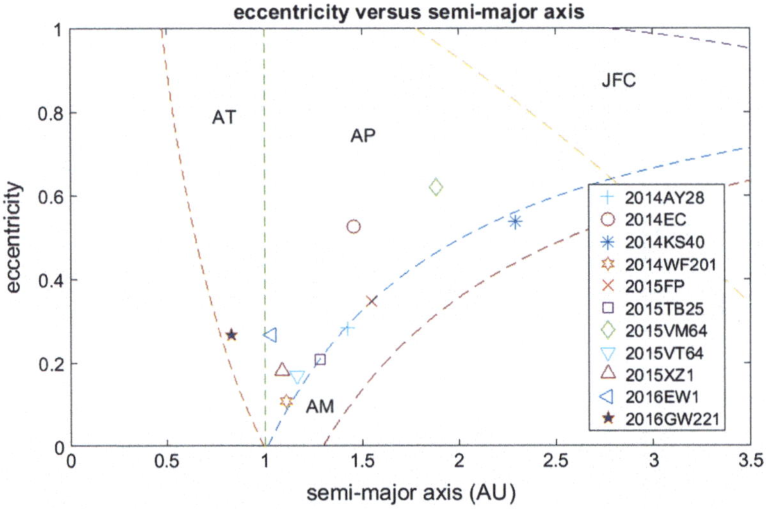


**Fig. 9** Eccentricity versus semi-major axis

Cgh, Ch) of the C-complex class. In the (a, e) space, 4 of them (2014 AY28, 2014 WF201, 2015 TB25, and 2015 VT64) (Fig. 9) are Amor asteroids with a < 2. They are among those Amors in the inner region of the main belt that Dandy et al. (2003) found mixed with E-type and M-type asteroids. 2015 VM64, and 2015 XZ1 are C-type asteroids found among Apollo asteroids. 2014 EC could not be classified with the two color plots or by comparing its reflectance with observed spectra; however, it can be ruled out as S-type asteroid since it is an Apollo asteroid with a > 2 (Dandy et al. 2003). 2016 EW1 was classified either as C-type or S-type. Being in the region of Apollo asteroids and having a < 1.5, it would be more S-type than C-type.

## Conclusion

The study of NEOs cannot be limited only to astrometry. It needs to be extended in the direction of physical characterization to know what NEOs are made of (their composition, and their structure). The more accurate method to accomplish this task is spectroscopy (optical and infrared). But being time consuming, and also expensive, the same task can be done with less expensive method: optical photometry, with a very inexpensive set of broadband filters (BVRI).

We were able to determine the color of 11 NEOs by combining different approaches. The first was the two color plots (Zellner et al. 1985) from color indices generated through photometry technique. The second approach was the comparison of NEOs relative reflectance normalized to V with the observed spectra in database. This second complements the first. Using the (a, e) space where different regions are populated by different asteroid families (Apollo, Amor, and Aten), and knowing the relationship that exists between these families and different colors of asteroids (Dandy et al. 2003), we were able to characterize the 11 NEOs observed with a 1.8 m telescope using Johnson-cousin optical filters.

We found that among the 11 NEOs, 7 are C-type asteroids or C-complex asteroids (Cg, Cgh, and Ch). There are 2014 AY28, 2014 WF201, 2015 TB25, 2015 VM64, 2015 VT64, 2015 XZ1, and 2016 GW221. Two could not be classified after the first approach, 2014 EC and 2015 FP. 2014 EC, being an Apollo asteroid with a > 2, is more likely to be S-type. 2015 FP is an Amor asteroid, and is a C-type asteroid. In (a, e) space, 2014 KS40 falls in the region of Amor asteroids and has a > 2. It is a probably a C-type NEO. Finally, 2016 EW1, classified either C or S in the two color plots, turns to be more S-type than C-type as it is an asteroid of Apollo family with a < 1.5.

## References


Bottke, W. F., Morbidelli, A., Jedicke, R., Petit, J. M., Levison, H. F., Michel, P., et al. (2002). *Icarus, 156,* 399.

Bus, S. J., Binzel, R. P. (2002a). *Icarus, 158*, 106.

Bus, S. J., Binzel, R. P. (2002b). *Icarus, 158*, 146.

Dandy, C. L., Fitzsimmons, A., Collander-Brown, S. J., Asher, D., Bailey, M. E. (2002). *Proceedings of Asteroids, Comets, Meteors*. ACM.

Dandy, C. L., Fitzsimmons, A., & Collander-Brown, C. J. (2003). *Icarus, 163,* 363.

Fedorets, G., Granvik, M., & Jedicke, R. (2017). *Icarus, 285,* 83.

Froeschle, Ch., Hahn, G., Gonezi, R., Morbidelli, A., & Farinella, P. (1995). *Icarus, 117,* 45.

Landolt, A. U. (1992). *AJ, 104*, 340.

Popescu, M., Birlan, M., & Nedelcu, D. A. (2012). *A&A, 544,* A130.

Popescu, M., Licandro, J., Morate, D., de Leon, J., Nedelcu, D. A., Rebolo, R., et al. (2016). *A&A, 591,* A115.

Rabinowitz, D. L. (1997). *Icarus, 127,* 33.

Rabinowitz, D. L. (1998). *Icarus, 134,* 342.

Schunova-Lilly, E., Jedicke, R., Veres, P., Denneau, L., & Wainscoat, R. J. (2017). *Icarus, 284,* 114.

Tody, D. (1986). The IRAF data reduction and analysis system. In D. L. Crawford (Ed.), *Proceedings of SPIE Instrumentation in Astronomy VI* (Vol. 627, p. 733).

Tody, D. (1993). IRAF in the nineties. In R. J. Hanisch, R. J. V. Brissenden, & J. Barnes (Eds.), *Astronomical data analysis software and systems II, A.S.P. Conference Series* (Vol. 52, p. 173).

Tricarico, P. (2017). *Icarus, 284,* 416.

Yoshida, F., Dermawan, B., Ito, T., Sawabwe, Y., Haji, M., Saito, R., et al. (2004). *PASJ, 56,* 1105.

Zellner, B., Tholen, D. J., & Tedesco, E. F. (1985). *Icarus, 61,* 355.

# Part II
# Stellar Evolution and Stars

# The Enigmatic Lambda Boötis Stars

**Christopher J. Corbally, S.J.**

**Abstract** The over 70 year story of λ Boötis-type stars does not have an ending, because we do not yet understand what makes them peculiar, but there is a lot we have learnt to date. The scientific journey takes us through the development of spectral classification of stars. Using spectroscopy, we know λ Boos' surface composition, which is normal in abundances relative to the Sun for carbon, nitrogen, oxygen, and sulphur, but underabundant for iron and similar elements. We gain a good idea of their relation to the Milky Way galaxy and their phase in the evolutionary stages of stars. We discover something of their relevance to the formation of planets around a star, and that a star continues to interact with material like gas and dust surrounding it and even with comets. We learn to treat simple solutions with suspicion, to assess where we have reached in our understanding of a scientific puzzle, and to keep hope that new techniques will bring future resolution to the story.

## Beginning an Unfinished Story

If you are among those who like a happy ending to a story, just turn to the next chapter. I quite understand. But if you enjoy the journey itself, in this case a scientific journey, with a challenge at the end, please read on.

The story really begins with a strong connection to the Vatican Observatory. Father Angelo Secchi, a Jesuit, was director of the observatory in the

Prepared April 2016, updated February 2017, for the book following the Symposium for the 80th Anniversary of the Vatican Observatory's move to Castel Gandolfo, September 2015.

C. J. Corbally, S.J. (✉)
Vatican Observatory, Albano Laziale, Italy
e-mail: corbally@as.arizona.edu

 
G. Gionti, S.J. and J.-B. Kikwaya Eluo, S.J. (eds.), *The Vatican Observatory, Castel Gandolfo: 80th Anniversary Celebration*, Astrophysics and Space Science Proceedings 51, https://doi.org/10.1007/978-3-319-67205-2_3

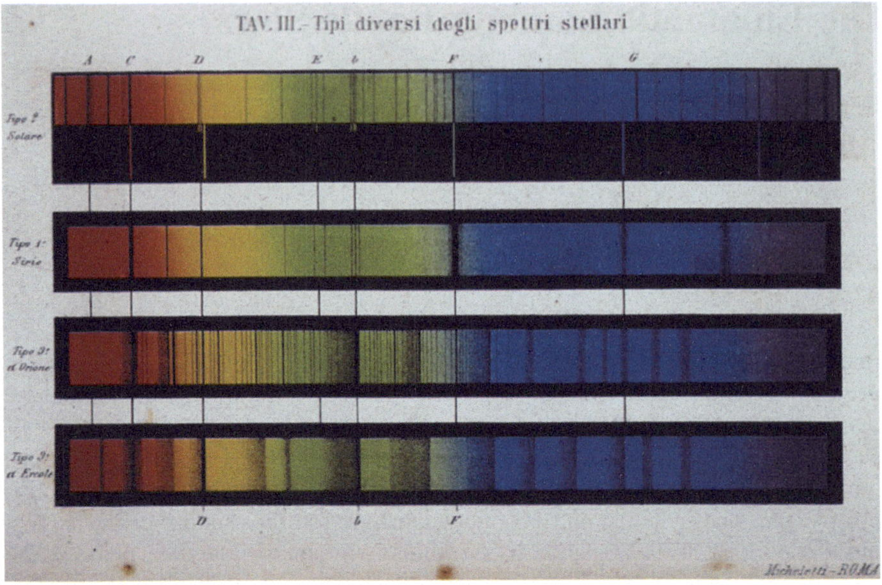


**Fig. 1** Angelo Secchi's drawings of different types of stellar spectra. The letters indicate features noted by Joseph Fraunhofer in the Sun's spectrum some 50 years earlier than Secchi's work

mid-nineteenth century. In the 1860s he was among the very first to observe stars with a spectrograph attached to a telescope, so extending their pinpoint light into a spectrum, or rainbow, one for each star.[1] He carefully drew what features he saw in the spectra, since this was the time before photography was available. Secchi was a prolific observer, eventually amassing spectrograms of some 4,000 stars. These were more than enough for him to notice that the spectra fell naturally into groups, or classes. His Type 1 were like the brightest star in the sky, Sirius, and Type 2 were like the spectrum of our own Sun. He later added three more types, but for now let us concentrate on Type 1 (Fig. 1).

The continuum of rainbow colors is interrupted by lines where light is absorbed in the atmospheres of the stars. Type 1, the second from the top, is represented here by Sirius. Type 2 on the top is the spectrum of the Sun. The bottom two spectra are both of Type 3, since both are characterized by broad, dark bands, later found to be due to Titanium Oxide.

The classification of spectra became more refined after Secchi's death, thanks considerably to work by a number of women at the Harvard Observatory, who were called quaintly for us now, "computers". They produced the Henry Draper catalogue, published between 1918 and 1924, giving spectroscopic classifications for 225,300 stars. A very significant refinement came in 1943 with the publishing of *An*

[1]For an account of Secchi and the early history of spectroscopy, see Gray and Corbally (2009, p. 2ff).

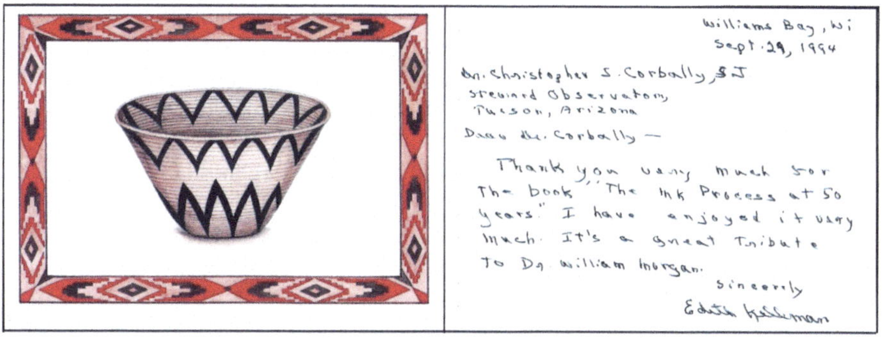


**Fig. 2** A Poma Basket "Thank you" card from Edith Kellman to Christopher Corbally

*Atlas of Stellar Spectra*, by W.W. Morgan, Philip C. Keenan, and Edith Kellman. That publication has a personal element for me. I had the delight to see Morgan and Keenan together in Toronto forty years later, just after I had finished my doctorate at its University; and I treasure a little card that I received in 1994 from Kellman (Fig. 2), who did the photographic work for the *Atlas*. She thanked me for sending her the proceedings of a conference celebrating the 50th anniversary of the "MK Process", as Morgan and Keenan's approach to classification became known (Corbally et al. 1994).

A note in the *MKK Atlas* to the page which treated stars like Secchi's Type 1, the A-type stars, gave a significant entry about a star whose spectrum didn't fit well into the normal sequence. "The spectral type of λ Boo is near A0, as far as can be determined. The spectral lines, while not unusually broad, are very weak, so that the only features easily visible are a weak K line and the Balmer series of hydrogen." λ Boötis was peculiar, even looking different to other peculiar A-type stars, so what was the reason?

## False Starts During Early Days

The story of finding out the source of λ Boo's peculiarity started off on a wrong foot. Some prominent astronomers in the 1950s claimed that the weakness of the spectral lines of this prototype star and others whose spectra were found to be similar, was just because they all belonged to what are called *Population II* stars. These are the oldest and most primitive stars in our Milky Way galaxy, formed at a time when our galaxy contained giant molecular clouds holding hydrogen and helium gases, and little else in the way of heavier elements. These heavier elements, called rather sweepingly *metals* by astronomers, built up to the kind of abundances seen in our Sun through many successive generations of massive stars synthesizing these metals through nuclear fusion. The massive stars then cast the new material out into the Milky Way through a dramatic explosion at the end of their lifetime, called a supernova event. That material was then added to giant molecular clouds from which came the next generation of stars. So, it took some 5 billion years for abundances of the metals to reach the concentrations that went into making our Sun. In contrast to the primitive Population II stars, stars like the Sun are called *Population I* stars.

When Oke (1967) studied the parameters of λ Boo stars in relation to stars like our Sun, he showed that they did indeed belong to Population I. So, we couldn't simply say that their lower metal abundances were just due to their belonging to the first generation of stars. Their abundances *should* be like the Sun's.

In the 1960s came further studies showing that while λ Boos were weak in metals, their oxygen was normal, i.e., like our Sun's oxygen abundance. Very puzzling. The puzzle was not helped by any A-type star which seemed not the usual run of "peculiar", being lumped together by observers with the λ Boos. The class became a garbage bag for the oddities among the A-type stars, and there are plenty of those.

## Addressing the Mess

With Richard Gray I shared the same thesis advisor at the University of Toronto, Dr. Robert Garrison, so we have longtime been friends and collaborators. For his thesis work in the 1980s Gray studied the whole gamut of A-type stars, and in doing so found that he could clearly distinguish the λ Boos by classifying their spectra in relation to the normal stars in the MK classification system. At last the "garbage bag" was becoming sorted and the true λ Boo stars could "shine" on their own (Gray 1997). You see, if a bag contains a whole variety of peculiar types, then isolating the mechanism that produces one such type becomes impossible. No wonder the problem of why this spectral oddity existed still was not solved forty years after W.W. Morgan had recognized it. (Philip Keenan concentrated on the cooler stars, amicably leaving the hotter ones to Morgan.)

In Fig. 3 we have something in digital form that Morgan would have examined via his photographic spectra. It is an example of λ Boo itself sandwiched in between the spectra of normal, or *standard stars*, for which it might be mistaken. This is Gray's own montage. The vertical axis shows the intensity of the light from a star, using a relative scale, against the wavelength (or color) of the light on the horizontal axis. Let us do a little spectral classification. (If you are not interested in the details, just skip to the next section for a continuation of the story.)

Above the spectrum of λ Boo is that of HR 3314, and β Leo is below. These are *early A-type stars*, or those with hot surface temperatures, around 10,000 K. The five deep and broad "teeth" are where light of a particular color is missing from the spectrum. They are due to the absorption by hydrogen atoms in the star's atmosphere of the complete spectrum of light coming from the star's interior. These features are called *spectral lines* from the stretched look of them on photograph plates and in drawings such as Secchi's (Fig. 1).

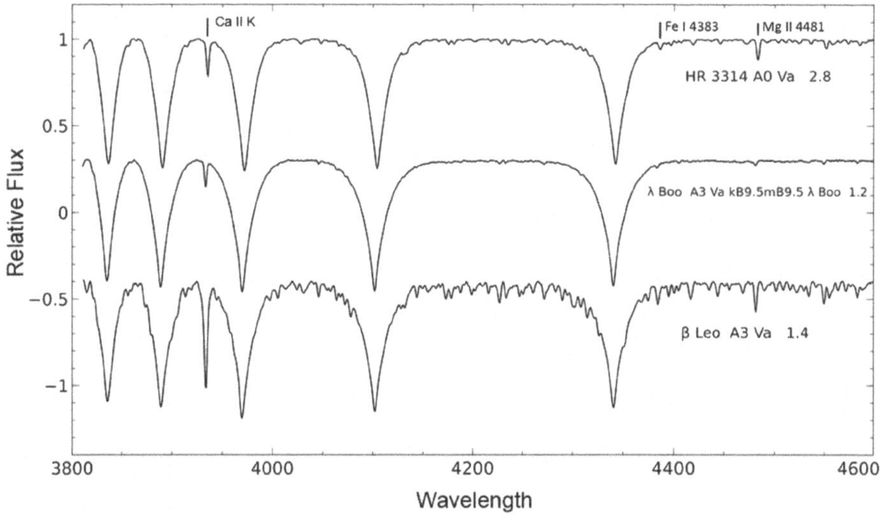


**Fig. 3** Montage of λ Boo with an MK standard star above and below. The spectra are from the Dark Sky Observatory

The spectral classification for λ Boo, A3 Va kB9.5 mB9.5 λ Boo, points to five things in its spectrum, all of which are important in defining a λ Boo star.

1. "A3" indicates that the strength of its deep hydrogen lines match those of the MK standard, β Leo. This is subtle, and best done by overlaying the spectra, so please take Gray's word for it! The A3 class refers to the surface temperature of the star, and λ Boos span A-type to early F-type stars.
2. "Va" says that it has an overall luminosity like that of a bright dwarf star, rather than like a giant star (class III), and certainly not a supergiant (class I). The last two classes belong to stars in their late stages of life. λ Boos are class V stars, fusing hydrogen to helium in their cores, like the Sun. They can also be referred to as *main sequence* stars. They have broad *wings* to their hydrogen lines.
3. "kB9.5" refers to the strength of the ionized calcium K line (*Ca II K*) on the left side of the spectrum. Compare it to the substantial Ca II K line in β Leo, to which it should match given its A3 temperature type, and to the still significant one in HR 3314. It is a bit weaker in λ Boo even than A0, and if we took a look at late-B type stars, it would fit with a B9.5 spectrum.
4. "mB9.5" is short for looking at the *metal* lines due to other elements, particularly iron. These are the tiny "wriggles" in the spectrum. They are weak, like the Ca II K line, again having about the strength matching a B9.5 star. A λ Boo star is weak-lined, at least regarding elements we see in this spectral region, compared to the strength of lines corresponding to its temperature type, in this case A3.
5. "λ Boo" is the final judgement that this star doesn't just have weak metal lines, but that the weakness is itself peculiar and imitates the prototype, λ Boo. A place to look for this is by comparing the ratio between the Mg II 4481 line

and the nearby Fe I 4383 line. That ratio (1.2) appears at the end of the classification in Fig. 3. In λ Boo the Mg II line is noticeably weaker than the iron lines nearby, when compared with the spectrum corresponding to its metal lines (HR 3314), and that is the real clincher.

There is another thing that one can tell from a spectrum. The broader the spectral lines are, the faster the star is rotating. This comes from the Doppler effect due to rotation. While the A-type stars have quite a range of rotational velocities, Gray found that λ Boo stars follow the regular distribution of these velocities. It had been claimed that the faster rotators tended to be λ Boos, which of course might have had a bearing on the reason for its peculiarity.

So we have a profile of what a λ Boo-type star looks like. Unfortunately, even since the time of Gray's definition, not everyone was as careful as he in classifying a star as a λ Boo type. The "garbage bag" was still being filled with non-members of the class. Simon Murphy, filled with enthusiasm from just finishing his doctorate, undertook the arduous task of searching the astronomical literature for stars that had been called λ Boos. He culled 212 stars and, together with eight collaborators, including Gray and myself, looked at them carefully. About 90 new spectra were observed for these, including some I took at the Vatican Advanced Technology Telescope on Mount Graham, Arizona. The re-evaluation of all 212 led to us settling on 64 stars being true members of the λ Boo class, 103 being definitely non-members, and a further 45 for which membership status is still unclear. The "unclear" category shows both that there can be marginal members of the class, and that despite efforts, in some cases we lack all the relevant information. This long paper was published near the end of 2015 in Publications of the Astronomical Society of Australia as *An evaluation of the membership probability of 212 lambda Boo stars*: *I. A Catalogue* (Murphy et al. 2015).

## The Exoplanet Connection and the Infrared

What is all the fuss about λ Boos? Can't we just leave them as an enigma and get on with other more important research topics?

Richard Gray and I don't think so. Since his thesis work, we've been trying to solve the puzzle of the mechanism behind their distinctive spectra. Even though they are a very small percentage of A-type stars, we feel that they hold a key to an aspect of the formation and/or evolution of all A-type stars.

In recent years, we have come together with other colleagues who think that this hidden and important aspect may relate to our understanding of star-planet interaction. Planets are made from the debris of gas and dust left over from the formation of their parent star. This appears like a great *dust ring* about the star, such as astronomers have been able to image around Fomalhaut (Fig. 4). Now, Fomalhaut (pronounced "foam-a-lot"!) is of type A3 Va, so an early-A star like the λ Boos,

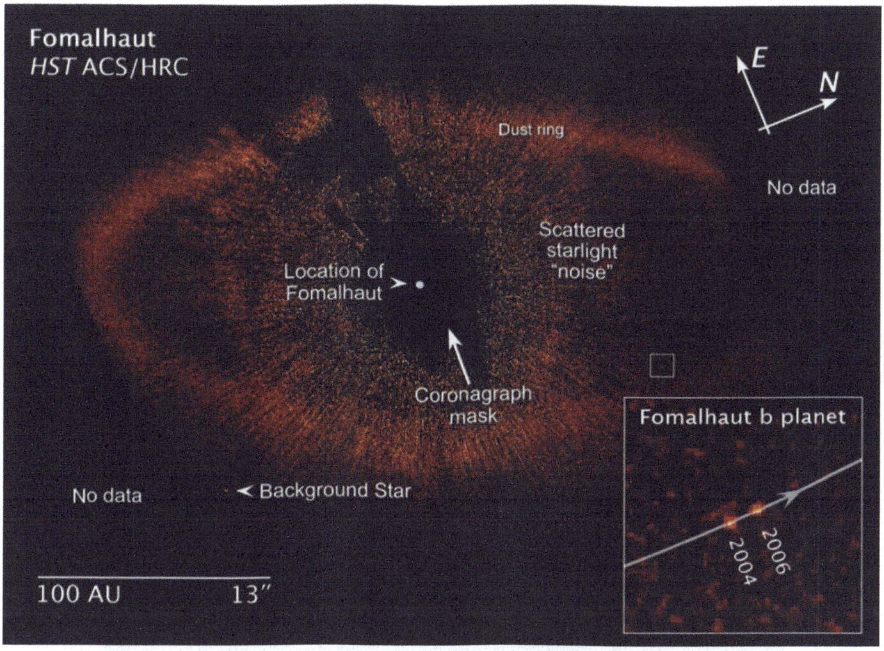


**Fig. 4** The dust ring and planet imaged (see inset) around the star Fomalhaut. NASA picture

and it has what most now accept to be an image of a planet embedded in a considerable ring at a good distance from the parent star. But is not a λ Boo star!

Another star, HR 8799, has not just one imaged planet, but four (Fig. 5). Significant to our argument, it is definitely a λ Boo star, with a classification of F0V kA5 mA5 λ Boo. This classification means that its temperature type is *later*, or cooler than λ Boo, and the weakness of its metal lines is just a little more. From other observations, it also has *excess infrared emission*, which means that it radiates in part of the infrared more than normal for a star of its surface temperature. Such excess is generally associated with a dust ring. Its ring is not surprising in view of the planets found around it, and indeed in 2009 the Spitzer Space Telescope imaged its dust ring in the infrared (Fig. 6). That enormous ring seems to be much stirred up, presumably from embedded planets and smaller objects, such as populate the outer parts of our Solar System. So, even though both have surrounding planets and dust disks, something is making HR 8799 different from Fomalhaut, at least as far as their atmospheres are concerned; and it is their atmospheres that give us their spectral characteristics.

This leads to another debate: do all λ Boo stars have dust (or debris) disks around them? Estimates have suggested, "not all", and ranged the ratio widely between 23% and 77% of A-type stars. A recent study, which tried to make the identification of λ Boo stars as unbiased as possible, provisionally puts the ratio at

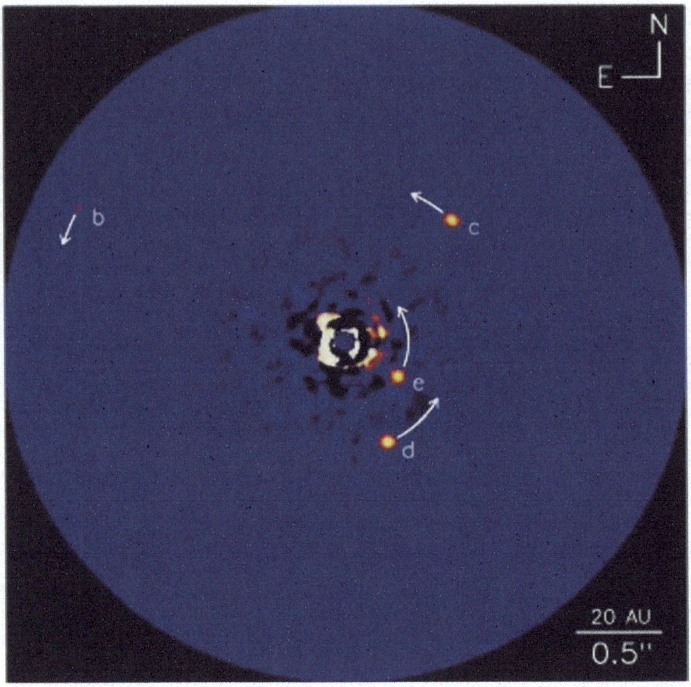


**Fig. 5** The four planets around the λ Boo-type star HR 8799, with arrows representing their motion over the coming 10 years. The light from the star has been blocked out. (*Credit* NRC-HIA, C. Marois, and Keck Observatory)

about 18%, which is close to the ratio of all A-type stars having infrared excess (about 15%, Gray et al. 2017). Yet given that, as we shall see a bit later, the phenomenon seems to be related to the selective accretion of material, i.e., gas and dust, onto the star's surface, how can diskless stars still have the λ Boo characteristics? Well, the low detection rate of infrared excess may just be an observational bias. It is reasonable to expect that the further flung, fainter λ Boo stars may also have infrared excesses, and so dust disks, when more sensitive data is available. If they don't and the nearer ones don't as well, that too could be relevant to the connection between the elusive "lambda Boo-ness" phenomenon and the mechanisms of dust disks, planet formation, and constant impacts of comets on their stars. It may be connected to massive Jupiter-like planets that are migrating in and out with respect to their planetary system, so affecting the distribution of gas and dust near their parent stars. The possible mechanisms are multiple, intricate, and dependent on the age of the star-planets system. We'll just have to be patient in sorting out all the possibilities.

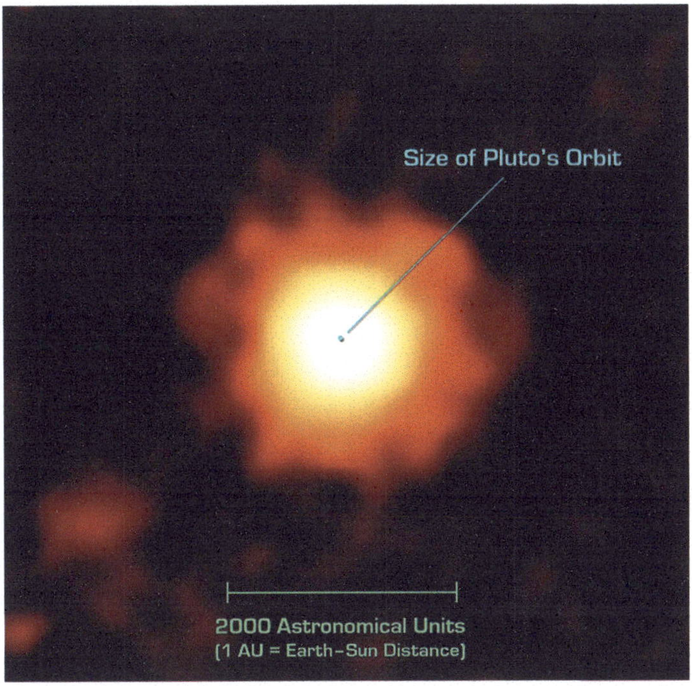


**Fig. 6** The enormous debris disk around the λ Boo-type star HR 8799. Note that the scale of this infrared image is very much smaller than in Fig. 5, so the disk extends way beyond the four planets. (*Credit* Kate Su, University of Arizona, from the NASA Spitzer Space Telescope)

## Ultraviolet Characteristics

There is a spectral region on the other side of the visible from the infrared, the ultraviolet or UV. Once this kind of observations came available for a significant number of stars via the International Ultraviolet Explorer (IUE), the comparison of λ Boos with normal stars could begin in the UV. Figure 7 gives an example of this comparison, and it is dramatic how in the λ Boo-type star, 29 Cyg, the aluminum line at the 1670 Angstroms (Å) wavelength is significantly weaker than the carbon. Indeed, carbon in this λ Boo, from its line at 1931 Å, is pretty much normal for an A7 type star, whether a dwarf (V) or giant (III). You will also notice that the fluxes of the λ Boo star are much higher in this region than those of the two normal stars. This is also found in stars which are metal-weak compared with normal stars, not just λ Boos. It is due to their lacking the strong metal lines which *blanket* the flux, making the overall flux weaker in normal stars. So, it is the ratio of Al II to C I, combined with the high overall flux, that tells us this λ Boo star is not only metal weak, but selectively so, having elements such as carbon in "normal", or near *solar abundance*.

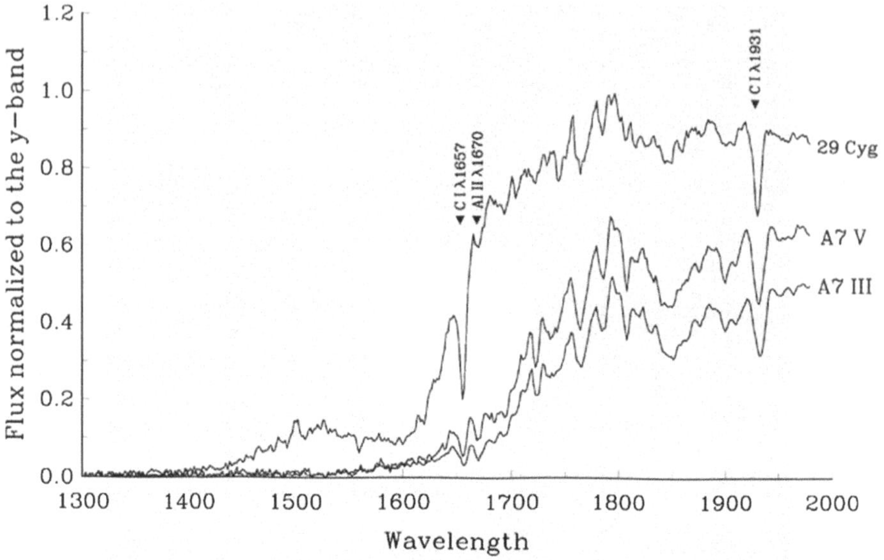


**Fig. 7** C I/Al II ratios in the ultraviolet for 29 Cyg, an A0.5 Va λ Boo star, compared with A7 V and A7 III standard stars. The montage was made up from IUE spectra by Richard Gray (Gray and Corbally 2009, 198)

Recently, the UV has been yielding further important information. Kwang-Ping Cheng ("KP") and James Neff have been mobilizing a group of students[2] to make a detailed study of ultraviolet spectra for stars with the full range of surface temperatures and other criteria that λ Boo-types show. They asked Gray and me to join the project. The idea, according to Neff, was to carve out an exclusive subset of stars that met every possible λ Boo criterion, and to focus on them for detailed follow-up studies. Those studies began by matching the ultraviolet spectra from the IUE and from the Hubble Space Telescope for this subset with modeled spectra from Gray's code, SPECTRUM, computed using appropriate physical properties for each of the stars. These properties include the detailed abundances of elements in the atmospheres of these stars, their surface temperatures, the *gravity* or compactness of their atmospheres, and their *microturbulent velocity*. The last property refers to the variation over small distance scales of the gas velocity, produced by turbulent motions of the gas in the outer layer of a star. This outer layer, or atmosphere, is where the spectral features are produced, and the effect of microturbulence is selectively to broaden the strong spectral lines versus the weak ones.

When the modeling is done right, so that the model spectrum matches as closely as possible the actual ultraviolet spectrum of the λ Boo star, then one has a value for the abundance of each element that the spectrum shows. For a given spectral line,

[2]These were Dustin Johnson, Erik Tarbell, Christopher Romo, Arvind Prabhaker, and Patricia Steele.

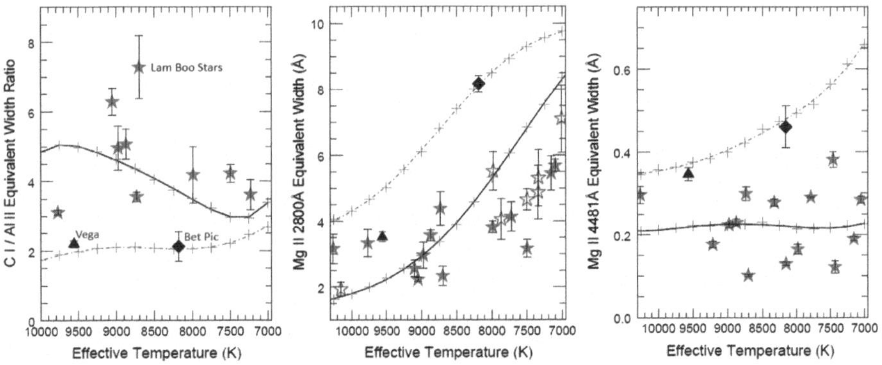


**Fig. 8** UV and Visible (the rightmost plot) characteristics of λ Boos, Bet Pic (diamond) and Vega (triangle). The solid line indicates a "mean λ Boo" model by Heiter (2002), and the dashed-dotted line indicates models with solar abundances (Cheng et al. 2016)

the abundance of the element that produces it determines the *equivalent width* of that line—low abundance gives a low equivalent width, and vice versa. Also, ratios of equivalent widths for pairs of elements can show the relative abundance of those elements. KP and colleagues plotted these values and ratios (Fig. 8). They also plotted the curves for a "mean abundance" λ Boo star and for normal A-type stars whose abundances are like the Sun.

It is clear that the abundances of magnesium and the ratios C I/Al II are different for λ Boo stars and their models, compared with normal stars. The λ Boo stars do form a clear subset, as was expected. What is also clear is that the stars Vega (the triangle) and bet Pic (the diamond) are not of λ Boo type from their abundances. This confirms the MK classification of their spectra in the visible as "normal", though some thought that they might be λ Boos. So, we do have a good way of supplementing MK classification in confirming the λ Boo status of stars. Unfortunately, at the moment we do not have UV spectra for all the 64 stars that were designated as λ Boo by Murphy et al. (2015), and the prospect of getting more orbiting ultraviolet telescopes is not promising.

## Abundance Pattern of λ Boos

There was a very significant study done in the late 1980s by Kim Venn for her doctoral degree at the University of Texas, Austin. She had an expert in stellar atmospheres, David Lambert, as her supervisor. He had proposed Venn look into the detailed pattern of element abundances in three λ Boos and compare these with the normal, solar abundances. She found that the λ Boos had severe underabundances of iron, magnesium, calcium, titanium, and strontium (the high melting point or *refractory* elements), while the abundances of carbon, nitrogen, oxygen, and sulphur (the *volatile* elements) were normal compared with the sun. In the case

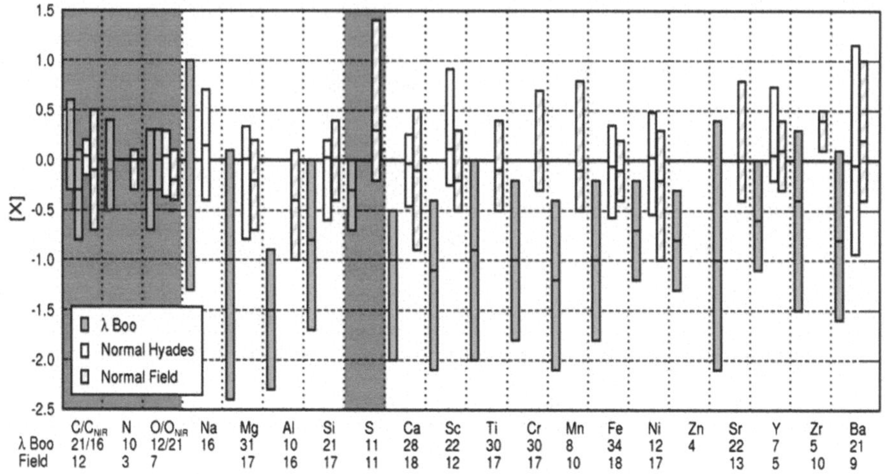


**Fig. 9** Abundance pattern of the elements for λ Boo stars, plotted as a range of values. The shaded elements, C–N–O–S, have near normal abundances, whereas others can range to quite low values. Plot by Ernst Paunzen after Heiter (2002)

of aluminum and magnesium, with respect to carbon, we have just seen this effect in the ultraviolet. Venn and Lambert (1990) concluded that, since this abundance pattern resembled that found for gas between stars (*interstellar gas*) where the metals are depleted by going into the formation of interstellar grains, then λ Boo stars might occur when gas around the star (*circumstellar gas*) or the interstellar gas is separated from grains and accreted onto the surface of the star. In this scenario, we should see the signature in the infrared of the interstellar cloud or circumstellar disk. This links λ Boos up with the formation of exoplanets, which we touched on earlier.

Later, Heiter (2002) looked at the abundances of more λ Boos, increasing the sample to 50% of the known class, and confirmed and expanded the pattern to include more elements (Fig. 9). There is something distinctly odd about the abundances of λ Boos compared with normal stars. Either this "something odd" extends throughout the star, and so was acquired during the formation of the star, or it has the nature of a "skin disease", accreted later and affecting only a few of the A-type dwarf stars for a limited time.

The skin-disease designation has gained support since Oke found in 1967 that λ Boos are of the same generation as the Sun in the Milky Way, Population I. They come from the same region of our galaxy as the Sun and share its general motion, and so the abundance pattern of their interior composition should be just like that of the Sun.

With this in mind, Kamp and Paunzen (2002) took seriously a couple of facts about λ Boos. First, they comprise just 2% of the A-type stars in the Galaxy's field, i.e., those outside of clusters of stars. Secondly, their abundances are, as we have just seen, similar to that of interstellar gas. Maybe what we are finding on their surfaces is

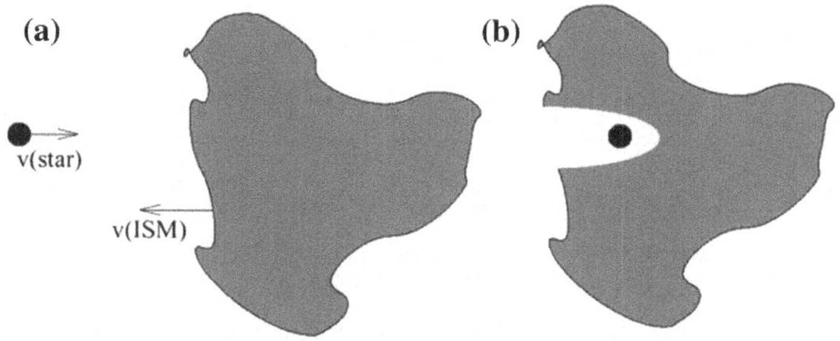


**Fig. 10** Sketch of a star **a** encountering a diffuse ISM cloud, and **b** clearing a cavity inside as it accretes material onto its surface (Kamp and Paunzen 2002)

the result of the star recently passing through a diffuse interstellar cloud (Fig. 10). That cloud would have to be small, but greater than 1 parsec (= 3.26 light-years) in diameter to allow sufficient accretion of its gas to produce the λ Boo effect. Such small interstellar clouds are presumed to be abundant, even though difficult to detect. After passage through the cloud, the persistence of the λ Boo effect would last only a matter of a million years, and then the star would revert to having a normal spectrum. This "reversion" would keep the number of λ Boos down to the observed small percentage in the field. This scenario explains a further fact. There are four pairs of stars, whose two components circulate each other by their mutual gravity, and both components of the four binaries show λ Boo characteristics. Could these have been acquired by a pair's simultaneous passage through an interstellar cloud? Maybe, but there are theoretical difficulties in the selective accretion of gaseous material onto the surface of a star during its passage through a cloud, while leaving aside the refractory material. It seems that the gas just won't "stick" onto the star in enough quantity!

## Age and Environment: Or Looking for the Right Accretion Parameters

In the puzzle about a mechanism to produce the λ Boo effect, the problem with the diffuse cloud scenario puts us back to looking for a solution at the immediate disk of gas and dust surrounding the star. Gray and I understood that this disk would be closest to the star and densest soon after the formation of the star from its original "cocoon" of gas and dust. Presumably it would be then, when the star was young, that the exchange between this disk and the star was going to be the most active. It would leave the more refractory elements out in the disk, while the more volatile carbon, nitrogen, and oxygen were acquired back on to the surface of the star. The λ Boo effect should decline with age, reducing their percentage among A-type stars as a whole. So, we observed spectra for stars that were good λ Boo candidates first in stellar associations of a few million years old, and then in intermediate age clusters.

The latter had ages ranging from 30 million to 1 billion years, very similar to the ages of A-type stars found in the field.

The results were curious. In the very young associations we found one *bona fide* λ Boo star and one marginal one. This put the percentage of them at 2% of the A-type stars in the associations, the same as in the field. However, for the intermediate age stars in the clusters, 130 stars, we could not find a single λ Boo star. (As a reassurance of our search method, we did find two λ Boos, but determined later that they were not cluster members.) Maybe we were just unlucky. We increased our sample by a further 184 A-type stars in clusters whose spectra had been classified by the most reliable authors. Again, none were found. This was a significant *null result* since there was only one chance in a hundred that we could have been so unlucky as to find no λ Boo stars of intermediate age in clusters. We concluded that there was something special about the environment for a star in a cluster compared with that in the field. In a cluster, stars may be close enough together to reduce the persistence of the material immediately surrounding any star. This reduction in material could be caused by the intense radiation from nearby, massive bright stars. In the outer parts of a cluster, stars are more loosely bound by gravity to the cluster and so can escape it readily. Also, the radiation from those massive stars would be less. This would mean that the primitive disk surrounding these escaping stars could persist when they became field stars, some with enough of the material to become and remain as λ Boos for a time.

There are other modifications of this simple scenario. A promising one uses comets falling into a star, as they are observed to do for our Sun, to provide a continuous, volatile-element rich deposit onto its atmosphere. Or, for very young stars still with substantial disks near them, large Jupiter-like planets that are orbiting close to their star, the so-called *hot Jupiters*, could keep the refractory dust at a distance while allowing the volatile gas to spiral onto the star's surface (Fig. 11). Yet another looks to a dual mechanism: young λ Boos are produced by the accretion of volatiles from the circumstellar disk; older λ Boos have had time for *diffusion* to raise the light, volatile elements in a star's atmosphere by the strong

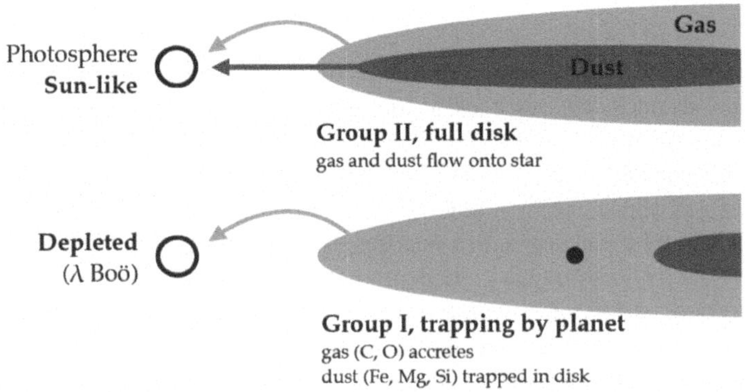


**Fig. 11** Two types of young circumstellar disks with and without a nearby giant planet. The giant planet in the lower disk separates the volatile gas, which accretes onto the star, from the dust which remains in the disk, a good distance from the star (Kama et al. 2015)

radiation coming from its interior, while the heavier elements have sunk below the point in their atmospheres where we can see them in our spectra. The older λ Boos could also be accreting from material blown out by the hot Jupiter winds. However, none of these mechanisms satisfactorily explains why we don't see λ Boos in intermediate-age clusters. So as yet, no theory has compelling evidence for it.

## The Binary Hypothesis, or the Simple Solution

Wouldn't it be much simpler if the explanation for the appearance of λ Boos was because in their spectrum, instead of just one star, we are really seeing a combination of two stars, orbiting so close that we can't distinguish them individually? This idea was very attractive to two colleagues, Rosanna Faraggiana and Michèle Gerbaldi, who were joined by other collaborators. Gray and I had long respected the work of these two, but we didn't believe their binary hypothesis for λ Boos. Both of us had experimented with combining the spectra of two normal stars, Gray numerically and I somewhat earlier, photographically. While we could produce *composite spectra* showing weakened lines compared with a single normal star, they did not have the full-fledged characteristics of λ Boos. Also, there are many visually-close binaries in clusters, so we should have seen at least a few λ Boos in our sample of cluster-star spectra.

There was only one way really to settle the issue. At a meeting on A-type stars in Poprad, Slovakia, when this "just a binary" hypothesis was floating around, I challenged a friend, Elizabeth Griffin, who had regular access to the Dominion Astrophysical Observatory's 1.2 m telescope in Victoria, to make definitive observations. Its spectrograph could show features at high resolution, sufficient to detect if the spectral lines were shifting very slightly to the red or the blue as a result of orbiting another, unseen star (a shift called *the Doppler effect*). Griffin took on 12 stars, which had been listed by Faraggiana and Gerbaldi as having "variable radial velocity", and observed their spectra exhaustively over 6 years at deliberately irregular intervals. Three of these stars showed small variations in the profile of some spectral lines, but none of the variations could be ascribed to the presence of a companion star (Griffin et al. 2012). The simple solution was out. Rather like sorting out the genuine λ Boos from the just peculiar A-stars that had been thrown together into the original garbage bag, this second null result took a lot of steady effort; but it needed doing if the real mechanism for λ Boos would ever emerge.

## Status of the Enigma

We have actually come quite a long way in the scientific journey we have been making.

- We have found that λ Boos are distinct spectroscopically over a particular range of stellar surface temperatures, those associated with the spectral types B9 to F3.
- They are not due to double stars whose two spectra we cannot distinguish, the composite-spectrum binaries.
- They age from very young stars, before they have reached the main sequence, up to stars at the end of their hydrogen-fusing stage.
- They are not found in clusters after ages when A-type stars have reached the main sequence.
- They are found evenly distributed throughout the field stars in our galaxy, of which they compose 2% of A-type stars there. That is also their frequency among very young stars in clusters.
- While we know that they have peculiar atmospheres, they may also have near solar abundances in their interiors (see below); but if the peculiar abundance pattern extends down through the star, this would radically modify our ideas of how 2% of A-type stars form within nebulae.
- They seem due to a selective accretion of gas and volatile elements onto the star's surface, while the more refractory, *iron-peak elements*, stay surrounding the star, away from its surface.
- Diffusion, accretion, rotation, and pulsation have all been considered as mechanisms. Accretion looks the dominant one, but the others may be involved depending on the age of the λ Boo.
- The accretion may well be triggered, at least for a while, by hot Jupiter planets close to the star interacting with its debris disk.

I hope you share the sense that solving the enigma of why we have λ Boos at all will lead us to understand better how dust and gas evolve in the region around stars. There is something peculiar to the λ Boo arrangement of star, debris disk, and planets that turns on the phenomenon for just a couple of percent of the A-type dwarf stars. In the scenario of material around stars we do not fully understand how elements combine together into bigger grains, which will then form the smaller, and finally larger bodies in a planetary system. So the whole enigma of λ Boos probably involves an important contemporary question: what is the correct mechanism by which planets are formed around their parent stars. This is not the time to stop trying to understand λ Boos!

## Lessons from the Enigma

Let us reflect on our scientific journey. The ultimate reason for the unique spectra of λ Boos still remains an enigma. But the journey so far has been fun to do, has involved fine colleagues, has made progress, and contains good lessons on how to get answers to questions, and not just scientific ones. These lessons have been:

- *Isolate the phenomenon*
  This was when, some forty years after the MKK Atlas pointed out the special spectrum of λ Boo itself, Gray clearly isolated their characteristics from other peculiar A-type stars.
- *Get a clean sample*
  Cleaning up the garbage bag so that just the genuine λ Boos remain, is tedious but has to be done. Kudos to Murphy for addressing the mess by a thorough search though the literature for any star that was called "λ Boo", and for leading the evaluation of these.
- *Expand the available information*
  Though the phenomenon of λ Boos was first spotted in the blue-green section of the visible spectrum, it is important to find whether it shows up in other parts of the electromagnetic spectrum. The exploration of the ultraviolet characteristics of λ Boos is proving very fruitful, thanks recently to Cheng, Neff, and collaborators. So too, the excess radiation in the infrared of all four nearby λ Boos and some of the more distant ones, is an essential indicator of the dusty and gaseous disks surrounding these stars.
- *Check the obvious*
  When puzzling over the reason for the peculiarities, some overall characteristics are obviously needed: what are the composition differences between λ Boos and normal stars; for what ages of stars does the phenomenon occur; where are they found in the galaxy generally and in specific star forming locations?
- *Don't be afraid of a null result*
  A null result, while not making the lead in any news report, can be really helpful in eliminating false alleys, like the binary star hypothesis.
- *Amass the possibilities*
  Possibilities for an explanation have to be in the back of one's mind all the while that one makes new observations and does modeling. Sometimes it is very fruitful to sit back and check the possibilities off against findings, in order to discover the next step needed in the journey.
- *Keep hopeful!*
  Without hope, one would never start a scientific investigation, and never continue, despite null and frustrating results.

## Future Research Lines

This review of the unfinished story will stimulate ideas on what we need to do to give it a satisfactory ending. There are some very promising research lines with respect to λ Boos, all with challenges to complete them.

One important, lingering question is whether the λ Boo pattern of abundances is really just a skin disease, affecting its atmosphere only, or whether the whole of the λ Boo star has this unusual abundance. Happily, there is a quite promising, new way to find out, *asteroseismology*. Just as sensors on the surface of the Earth can pick up the tremors produced by earthquakes, and these allow us to model what the interior of the Earth must be like, so we can do the same with "star-quakes". Through identifying their frequencies, we can find the internal structure of stars, which will include the relative abundances of elements. Now, we obviously cannot put our seismic instruments on the surface of a star, but we can observe the effect of a star's "quakes" on the output of light from its surface. All stars, including our Sun, quake (*pulsate*) to some degree, but some λ Boos pulsate quite considerably. These belong to the delta Scuti class of pulsators (δ Sct), with varying *p-mode*, or pressure waves across their surfaces. A few of these delta Scuti-type λ Boos are also gamma Doradus pulsators (γ Dor), whose *g-modes*, or gravity modes, propagate from deep in the interior of the star. They are called *hybrid pulsators* since they have two modes of pulsation. Fortunately, their two modes can tell us about their interior composition as well as their surface abundances. Asteroseismology has already been done on a very few such λ Boo stars, but I find that the results are not conclusive as to their interior abundances. These are highly dependent on the stellar models used and assumptions made. But we do expect these to be refined and bring us more reliable results. This is particularly important since the Kepler satellite telescope[3] has given us a small, but significant database of pulsating λ Boo-type stars on which to work (Corbally et al. 2016). The Transiting Exoplanet Survey Satellite (TESS) will considerably expand this database once launched.

Another promising research line is the study of debris disks around stars. People are getting images, even from ground-based telescopes, of these disks in the infrared, but these images show debris at some quite considerable distance from their stars. What we really need to know is the distribution and characteristic of debris close to the star, and for that we look to the James Webb Telescope, which will operate in space and have excellent resolution.

Adaptive optics are now allowing large ground-based telescopes to have remarkable resolution. If the accretion onto λ Boo stars is really coming from the ISM rather than its immediate, circumstellar environment, then adaptive-optics spectroscopy of stars that are close together and embedded in dense ISM regions, could demonstrate this. However, the bow wave as the gas in the cloud gets compressed by the star ploughing through it, would be hotter than the infrared emission from several λ Boo stars indicates (Draper et al. 2016). It seems that ISM accretion does not apply to them.

Of course, the ultraviolet remains a promising field to test models of λ Boos and to find signatures of gas and its motions around these stars. We are hoping for more observations of some λ Boos before the Hubble Space Telescope stops working after its already amazing lifespan. However, there is no obvious successor to its capabilities in the ultraviolet. Rather than end on a sour note, let us remember the last lesson point: keep hopeful!

---

[3]A NASA planet-finder spacecraft launched in 2009 that made asteroseismology studies of thousands of stars in its field before it was repurposed as *K2* after reaction wheel failure prevented it concentrating on a single field (Wiki on asteroseismology).

## Further Reading

James B. Kaler (2011). *Stars and their spectra: an introduction to the spectral sequence*. Cambridge: Cambridge University Press.

(This classic introduction is updated and expanded from the original 1989 edition.)

## References


Cheng, K.-P., Neff, J. E., Johnson, D. M., Tarbell, E. S., Romo, C. A., Prabhaker, A., et al. (2016). Utilizing synthetic UV spectra to explore the physical basis for the classification of Lambda Boo stars. *Astronomical Journal, 151,* 105–121.

Corbally, C. J., Gray, R. O., & Garrison, R. F. (Eds.). (1994). *The MK Process at 50 Years: A powerful tool for astrophysical insight.* San Francisco: Astronomical Society of the Pacific.

Corbally, C. J., Murphy, S. J., & Gray, R. O. (2016). *Investigating pulsating Lambda Boötis Stars.* Poster paper at Stars 2016 conference. http://www.star.uclan.ac.uk/stars2016/posters/Corbally.pdf. Accessed February 6, 2017.

Draper, Z. H., Matthews, B. C., Kennedy, G. M., Wyatt, M. C., Venn, K. A., & Sibthorpe, B. (2016). IR excesses around nearby lambda Boo stars are caused by debris discs rather than ISM bow waves. *MNRAS, 456,* 459–476.

Gray, R. O. (1997). The λ Boötis stars: A review. In A. G. David Philip, J. W. Liebert, & R. A. Saffer (Eds.), *The third conference of faint blue stars* (p. 237). Schenectady: L Davis Press.

Gray, R. O., & Corbally, C. J. (2009). *Stellar spectral classification.* Princeton: Princeton University Press.

Gray, R. O., Riggs, Q. S., Koen, C., Murphy, S. J., Newsome, I. M., Corbally, C. J., et al. (2017). The discovery of λ Bootis stars: The Southern Survey I. *Astronomical Journal, 154,* 31.

Griffin, R. E. M., Gray, R. O., & Corbally, C. J. (2012). The composite-spectrum binary hypothesis does not explain the Lambda Bootis stars. *Astronomy & Astrophysics, 547*(A8), 1–14.

Heiter, U. (2002). The abundance pattern of Lambda Bootis stars. *Astronomy & Astrophysics, 381,* 959–970.

Kama, M., Folsom, C. P., & Pinilla, P. (2015). Fingerprints of giant planets in the photospheres of Herbig stars. *Astronomy & Astrophysics, 582,* L10.

Kamp, I., & Paunzen, E. (2002). The λ Bootis phenomenon: Interaction between a star and a diffuse interstellar cloud. *Monthly Notices of the Royal Astronomical Society, 335,* L45–L49.

Murphy, S. J., Corbally, C. J., Gray, R. O., Cheng, K.-P., Neff, J. E., Koen, C., et al. (2015). An evaluation of the membership probability of 212 Lambda Boo stars: I A Catalogue. *Publications of the Astronomical Society of Australia, 32,* 36.

Oke, J. B. (1967). Effective temperatures and gravities of lambda Bootis stars. *The Astrophysical Journal, 150,* 513–519.

Venn, K. A., & Lambert, D. L. (1990). The chemical composition of three Lambda Bootis stars. *The Astrophysical Journal, 363,* 234–244.

# Mapping the Milky Way in the Near-IR: The Future of the VVV Survey

**Dante Minniti**

**Abstract** The VISTA Variables in the Via Lactea (VVV) is an ESO public survey that has been mapping the bulge and inner disk of the Milky Way in the near infrared for the past 6 years. Here we examine the scientific goals and some key results, as well as the legacy of the VVV survey. We also discuss the making of the first huge public image of the Galactic bulge (a 140 Gigabytes single image with 25,000 Megapixels), and how these data allowed us to measure directly the total bulge mass of the Milky Way bulge, that is twenty billion Solar masses. Finally, we present the plans of the recently approved VVVX, an extended survey of the Milky Way, that would map about 4% of the sky repeatedly in the near-IR, measuring more than 2 billion stars, producing a $\sim$1 Petabyte database (containing images, catalogues and maps) for the whole community to exploit.

## Introduction

The Vatican Observatory is involved in a large survey. This is the VISTA Variables in the Via Lactea, or VVV for short. Via Lactea is the Milky Way in Spanish. The VVV is a public near-infrared variability survey of the European Southern Observatory (ESO), that is sampling the Galactic bulge and an adjacent section of the Southern Galactic plane. As shown in Fig. 1, our survey covers crucial stellar populations of the Galactic bulge, including the Galactic center, as well as the Galactic disk with regions of intense star formation around its spiral arms (Minniti et al. 2010).

D. Minniti (✉)
Departamento de Física, Facultad de Ciencias Exactas, Universidad Andres Bello, Av. Fernandez Concha 700, Las Condes, Santiago, Chile
e-mail: dante@astrofisica.cl

D. Minniti
Instituto Milenio de Astrofísica, Santiago, Chile

D. Minniti
Vatican Observatory, V00120 Vatican City State, Italy



G. Gionti, S.J. and J.-B. Kikwaya Eluo, S.J. (eds.), *The Vatican Observatory, Castel Gandolfo: 80th Anniversary Celebration*, Astrophysics and Space Science Proceedings 51, https://doi.org/10.1007/978-3-319-67205-2_4

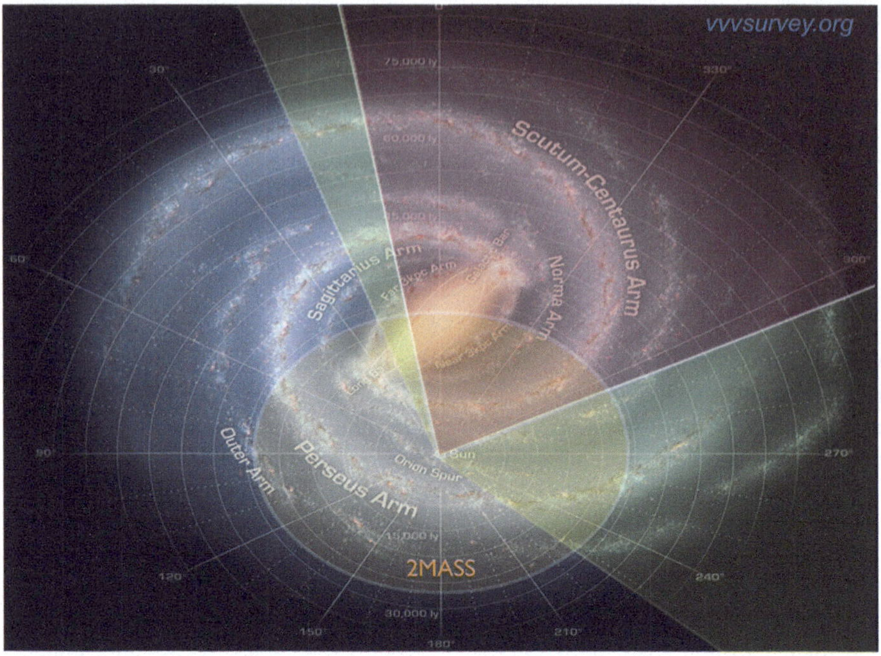


**Fig. 1** Limits of the IR surveys in the disk of the Milky Way. The transparent circle approximately shows how far can 2MASS see tracers like RR Lyrae variables or red clump giants. It turns out that the GLIMPSE survey with the Spitzer Space Telescope reaches similar depth in the Galactic plane. The transparent red wedge represents the depth of the VVV survey for these same tracers: they can be seen throughout the Galaxy and beyond. The transparent green extensions show the area to be covered by the future VVVX survey. The Galaxy representation shown in the background was made by Ron Kurz (JPL/NASA) based on the Spitzer GLIMPSE survey

The VVV Survey mapped more than 540 square degrees towards the central region of our galaxy using the 4 m VISTA telescope at ESO Paranal Observatory between the years 2010 and 2015. The VISTA telescope has a wide field camera, called VIRCAM, that is used for mapping the sky very efficiently in the near-IR (Emerson et al. 2004).

Our survey uses the 16 VIRCAM near-IR detectors in the ZYJHKs passbands to make individual images called paw prints. These paw prints are shifted appropriately in the sky to make mosaics that allow to have entire 1.5 square degree images (without gaps) called tiles. In the original VVV Survey we have 196 tiles in the bulge (tiles b201 to b396), covering a total region of 20 $\times$ 15 square degrees, plus 152 tiles in the southern Galactic disk (tiles d001 to d152), covering 4.5 $\times$ 55 degrees. The bulge images include the Galactic centre (located in tile b333), which is a very interesting region.

## Why a New Survey of the Milky Way?

We decided to make the VVV survey because the image of the Milky Way in the inner regions was not complete. These regions need to be studied in the near-IR, in order to overcome the heavy reddening and extinction across the Galactic plane. The previous massive near-IR survey of our galaxy is called Two Micron All Sky Survey (2MASS), made with observations in the late 90s, that produced many interesting discoveries (Skrutskie et al. 2006). In particular, the 2MASS all-sky map showed beautifully the Milky Way galaxy in the near-IR, being a realistic representation that we live in a spiral galaxy.

There are several differences between the 2MASS and the new VVV surveys. For example, 2MASS covered the whole sky using the JHKs filters, while VVV covers only 1.3% of the sky using the ZYJHKs filters. Also, the VVV has higher resolution, and reaches deeper magnitudes than 2MASS. Finally, VVV is a multiepoch survey, providing Ks-band light curves (with 50–300 epochs) for about one billion objects (Saito et al. 2012; Hempel et al. 2014).

There was another recent survey of the Milky Way disk and central regions with sufficient resolution to detect point sources in crowded fields: the GLIMPSE survey made with the Spitzer Space Telescope (Benjamin et al. 2005). GLIMPSE mapped the inner Galactic bulge (10 $\times$ 10 degrees) and plane (between Galactic latitudes of $-1$ and +1 degrees) using mid-IR filters.

Towards the Galactic centre in particular, extinction and source crowding did not allow the previous surveys to see very far. The VVV survey has higher resolution (1 pixel = 0.34 s of arc) and depth (down to Ks of about 17–18 magnitudes). This allows us to see all the way through the Milky Way, even in some very crowded and reddened fields of the bulge and disk. In other words, even though we knew that there was a forest because we saw the forest, just now we have the capability to see its innermost trees, to map them throughout the forest, and beyond.

Therefore, the main scientific goal of the VVV Survey it to unveil the three dimensional structure of the Milky Way using well characterised distance indicators such as RR Lyrae and Cepheid variables.

However, due to the large size of this survey, and the variety of interesting astrophysical objects found in the central Galactic regions, additional survey goals were defined by the science team, including to search for new open and globular star clusters, and to identify their variable stars; to map the star forming regions, variable young stellar objects (YSOs), and study other rare variable sources (X-ray binaries, cataclysmic variables, novae, and transients) across the Galactic plane. We would also search for microlensing events, for eclipsing binaries and for planetary transits; and identify red clump giants in the Galactic bulge and disk, and monitor the variability around the Galactic Center, and in the nearby dwarf spheroidal galaxy Sgr. Other projects are to find high proper motion objects, and to identify background QSOs that can be used for the astrometry; etc.

The VVV survey was planned to give the multicolor photometry of the whole area using the ZYJHKs filters during the first year. The following years were dedicated to the variability observations in the Ks-band only. Finally, the last year was dedicated to repeat the multicolor observations. In this way, we have multicolor observations with a baseline of about 5 years, and Ks-band light curves during that period of time.

The final product of the VVV Survey is a deep IR multiepoch and multi-band catalog in five passbands (from 0.9 to 2.5 μm), in order to make a census of variable objects and measure proper motions. Data from the survey are first processed at CASU, and then are made available to the entire community through the ESO Archive in Garching, Germany, and the VISTA Science Archive (VSA) facility in Edinburgh, United Kingdom. The VVV Survey webpage vvvsurvey.org lists additional information. The large VVV database (that would eventually contain well over 200 Terabytes) is fostering studies of the star and cluster evolution history of the Milky Way, and providing a population census of the Galactic bulge and center, in addition to star forming regions throughout the disk.

## The Story of a Giant Image: Entering the Era of Big Data

We wanted to see how the inner Milky Way really looks like by mapping a large sky area in the near-IR. These images can be used to obtain a global view of the bulge and study its constituent stellar populations. It was not easy, however, to obtain a good overall picture of the inner bulge. The mosaicing of each tile is just a first step, then we had to combine the different filters, correct for distortions matching sources, make the cross-calibration of the photometry, do the astrometry to produce an homogeneous map, correct for the overlapping tiles to make a seamless map, etc. All this for an enormous, gigantic database consisting of images and photometric tables. This amazing map was made by Ignacio Toledo in Chile, and is shown in a reduced format in Fig. 2.

We also wanted to distribute widely the image, and the help of the ESO Press Release department was needed for this. It is not easy to distribute such a large image, this was a pioneering work, the first of many huge astronomical images to be made available to the World. Indeed, it was the most downloaded image of ESO history, in spite of its large size (Christensen 2012).

Off course, this image was not only for press release. It contains about one billion stars, for which we have measured near-IR photometry. The aperture photometry is obtained at the Cambridge Astronomical Survey Unit (CASU), and is homogeneously calibrated throughout the VVV tiles (Irwin et al. 2004; Hambly et al. 2004), and additional point spread function (PSF) photometry is made using different pipelines (Alonso Garcia et al. 2015; Mauro et al. 2015; Contreras Ramos et al. 2017).

The data contained in the image of Fig. 2 allowed us for example to measure the total mass of the bulge (Valenti et al. 2016). We obtained this total mass by directly counting red clump giants using PSF photometry. However, this is not so simple as it sounds. Important corrections had to be made, including photometric

**Fig. 2** Real image of the Milky Way bulge made using the JHKs near- infrared filters. At full resolution (the scale is 1 pixel = 0.34s of arc), this is a 140 Gigabytes single image with 25,000 Megapixels. This giant image covers the central 300 square degrees of the Milky Way (oriented along Galactic coordinates, with longitudes from $-10°$ to $+10°$, and latitudes from $-10°$ to $+5°$). This was made of about 400,000 individual images, of $512 \times 512$ pixels each, by Ignacio Toledo

incompleteness due to source crowding, correction for large and variable extinction, source doubling in overlap areas, scaling to the bulge luminosity function, etc. The final result for the total mass of the Milky Way bulge is twenty billion Solar masses (Valenti et al. 2016), which is the first direct measurement of the mass of stellar sources and remnants in the Galactic bulge.

## The VVV Survey Legacy

The most important legacy of the VVV Survey is the homogeneous public database, that has yielded many interesting results. As specific examples of how the deep photometry and good spatial resolution of the VVV survey have already provided crucial mapping of the Milky Way we can highlight:

1. The red clump giants, which were employed to delineate the Galaxy's bar(s), structure, the bulge metallicity gradient, the total bulge mass, and the nature of

the extinction laws (Saito et al. 2011; Gonzalez et al. 2012, 2013; Valenti et al. 2016).
2. The RR Lyrae variable stars, which were used to measure the distance to the Galactic center and detect its old populations, and to map the structure of the bulge and inner halo regions (Dekany et al. 2013; Gran et al. 2016; Minniti et al. 2016).
3. The Cepheids and young stellar objects (YSOs) that are being used to map the star formation in the inner bulge, to find distant star clusters, to trace the spiral arms, and to identify new kinds of eruptive variable protostars (Dekany et al. 2015a, b; Contreras Pena et al. 2016).
4. The new star clusters: a couple of bulge globular clusters plus hundreds of open clusters that we discovered throughout the disk (Moni-Bidin et al. 2011; Minniti et al. 2011; Borissova et al. 2011, 2014; Barba et al. 2015).
5. The accurate astrometry that allowed us to make catalogs of high proper-motion objects, to discover new nearby brown dwarfs, and substellar companions (Beamin et al. 2013; Smith et al. 2015; Kurtev et al. 2016; Gromadsky et al. 2016).

The VVV Survey database will also feed large spectroscopic surveys of the Milky Way. This has already started to happen, with targets being provided for the GAIA ESO Spectroscopic Survey (GES, Rojas-Arriagada et al. 2014), and the GIRAFFE Inner Bulge Survey (GIBS, Zoccali et al. 2014).

## The Preparation for the Future: An Extended Survey

The VVV extension (VVVX) proposal was first drafted during the Vatican Observatory VVV Workshop on "A New Galactic Survey", held at Castel Gandolfo in May 2015, where more than twenty VVV Science Team members got together. Now that the VVV survey completed its data taking, there are still open questions, and we had to plan the next steps ahead. The main goal of the Vatican VVV Workshop was to elaborate a proposal for the VVVX. We needed to consider the synergy with other survey telescopes and cameras (like VST, LSST, DECAM, Gaia), and plan a legacy for the Astronomical Community of an extended and more complete survey of the inner galactic regions, which will be inaccessible for some of the other surveys. The VVVX will provide new targets and questions to address with additional dedicated spectroscopic survey instruments (like APOGEE-S, WINERED, MOONS, 4MOST), as well as large facilities to be built in Chile like the ESO Extremely Large Telescope (E-ELT) and the Giant Magellan Telescope (GMT), that can enable spectroscopic follow-up of interesting targets.

After much discussion (e.g. Fig. 3), we decided to map a total area of 1700 square degrees, obtaining multicolour near-IR photometry, as well as about 25 Ks-band epochs in the extended region. Also, the old VVV area will be re-observed for a few epochs in order to recover long term variability and to provide an extended baseline (of about 10 years) for proper motion measurements.

**Fig. 3** Discussion word map from the VVV Workshop, where the plans for the VVVX survey were made. The entire VVV Survey Science Team involves dozens of scientists from different countries, that cover different areas of expertise including Stellar Populations; Stellar Evolution; Star Clusters; Variable Stars, Distance Scale, Galactic Structure; Galaxy Formation; Gravitational Microlensing; Extrasolar Planets; etc

The VVVX proposal was successful (approved in 2016), with 200 nights awarded at the ESO VISTA telescope, and we are starting the first observations of the extended area shown in Fig. 4 (and also in Fig. 1). Even though VVVX will only cover about 4% of the sky, it will survey more than half of the Milky Way stars in the near-IR, because it is precisely mapping the densest regions of our Galaxy: the bulge and southern plane. This means, for example, that we have a fair chance of detecting a supernova in the Milky Way that explodes in the next few years.

We are well within the "big data" realm, with the VVVX public database expected to reach about 1 Petabyte (including advanced data products like images,

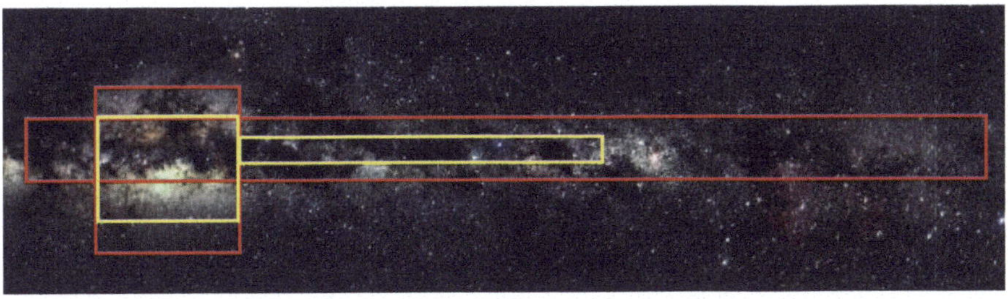

**Fig. 4** VVVX observing area projected on the optical picture of the Milky Way disk. The yellow region was already mapped by the VVV survey. The red region is the planned extension for the VVVX survey. This optical image is also oriented along Galactic coordinates, covering Galactic longitudes from −135° to +25°, and Galactic latitudes from 20° to +20° (Mellinger 2012)

catalogues, maps). With the VVV Survey we already obtained interesting scientific results, and with the extension VVVX there are many more to come. We dream about answering questions like: Are the oldest Galactic stars located in the central regions? What is the shape of the bulge? How are the different stellar populations distributed in the inner disk? And in the bulge-halo transition region? How many star clusters are missing? With the VVVX survey, we are discovering our own galaxy, fostering international collaborations, promoting Astrophysics at every level, and securing resources for the future generations.

**Acknowledgements** I am deeply thankful with the VVV Science Team for their superb work. We gratefully acknowledge data from the ESO Public Survey program ID 179.B-2002 taken with the VISTA telescope, and products from the Cambridge Astronomical Survey Unit (CASU). Support was provided by the BASAL Center for Astrophysics and Associated Technologies (CATA) through grant PFB-06, and the Ministry for the Economy, Development and Tourism, Programa Iniciativa Cientifica Milenio grant IC120009, awarded to the Millennium Institute of Astrophysics (MAS), and FONDECYT Regular grant No. 1130196. I am also grateful to the Vatican Observatory, and also to the Aspen Center for Physics, where this work was supported by National Science Foundation grant PHY-1066293, and by a grant from the Simons Foundation.

## References

Alonso-Garcia, J. et al. (2015). *AJ, 149*, 99.
Barba, R., et al. (2015). *Astronomy & Astrophysics, 581,* 120.
Beamin, J. C., et al. (2013). *Astronomy & Astrophysics, 557,* L8.
Benjamin, R. A., et al. (2005). *The Astrophysics Journal, 630,* L149.
Borissova, J., et al. (2011). *Astronomy & Astrophysics, 532,* 131.
Borissova, J., et al. (2014). *A&A, 569,* 24.
Christensen, L. L. (2012). ESO Outreach Com. Newsletter, Nov issue.
Contreras Pena, C., et al. (2016). *Monthly Notices of the Royal Astronomical Society, 465,* 3011.
Contreras Ramos, R., et al. (2017). *Astronomy & Astrophysics,* in press.
Dekany, I., et al. (2013). *Astronomy & Astrophysics, 776,* 19.
Dekany, I., et al. (2015a). *Astronomy & Astrophysics, 799,* L11.
Dekany, I., et al. (2015b). *Astronomy & Astrophysics, 812,* L29.
Emerson, J., et al. (2004). *Proceedings of the SPIE, 5493,* 401.
Gonzalez, O. A., et al. (2012). *Astronomy & Astrophysics, 543,* 13.
Gonzalez, O. A., et al. (2013). *Astronomy & Astrophysics, 552,* 110.
Gran, F., et al. (2016). *Astronomy & Astrophysics, 591,* 145.
Gromadzki, M., et al. (2016). *ActaA, 66,* 293.
Hambly, N., et al. (2004). *Proceedings of the SPIE, 5493,* 423.
Hempel, M., et al. (2014). *The Messenger, 155,* 29.
Irwin, M., et al. (2004). *Proceedings of the SPIE, 5493,* 411.
Kurtev, R. et al. (2016). *Monthly Notices of the Royal Astronomical Society, 464,* 1247.
Mauro, F., et al. (2015). *ASPC, 491,* 83.
Mellinger, A. (2012). arXiv:0908.4360.
Minniti, D., Lucas, P. W., et al. (2010). *New Astronomy, 15,* 433.
Minniti, D., et al. (2011). *Astronomy & Astrophysics, 527,* 81.
Minniti, D., et al. (2016). *The Astrophysics Journal, 830,* L14.
Moni-Bidin, C., et al. (2011). *Astronomy & Astrophysics, 535,* 33.
Rojas-Arriagada, A., et al. (2014). *Astronomy & Astrophysics, 569,* 103.

Saito, R. K., et al. (2011). *AJ, 142*, 76.
Saito, R. K., et al. (2012). *Astronomy & Astrophysics, 537,* 107.
Skrutskie, M. F., et al. (2006). *AJ, 131*, 1163.
Smith, L., et al. (2015). *Monthly Notices of the Royal Astronomical Society, 454,* 4476.
Valenti, E., et al. (2016). *Astronomy & Astrophysics, 587,* L6.
Zoccali, M., et al. (2014). *Astronomy & Astrophysics, 562,* 66.

# Some Recent Investigations in Stellar Evolution at the Specola Vaticana

**David Brown**

**Abstract** Studies in astronomy and stellar astrophysics have played a significant role in the ongoing research of the Specola Vaticana throughout its history. Within the last 150 years, this has been evidenced by the work of Angelo Secchi in the 19th century, by the Specola's participation in the 19th–20th century *Carte du Ciel* project, and by the Observatory's sponsorship of the prominent *Vatican Conference on Stellar Populations* in 1957. The latter was a milestone in the understanding of stellar populations. Since then, the astronomy community's knowledge of stellar populations, especially with regard to the various components of the Milky Way Galaxy and other galaxies, along with globular and open clusters, has increased greatly, commensurate with the advent of more powerful telescopes. The Specola Vaticana continues to contribute actively to such astrophysical studies by research conducted on photometric surveys of portions of the Milky Way Galaxy and also by the study of extreme horizontal branch stars that are thought to be ubiquitous in most stellar populations.

**Keywords** Stellar populations · Extreme horizontal branch stars · Clusters

## Stellar Astrophysics and the Specola

At the crossroads between the immense cosmos that confronts humanity on a large scale and the ordered and infinitesimally small reality (micro-cosmos) that underlies it in the atomic/quantum realm, stand the stars, subject to the same laws of physics that govern both the micro-cosmos and macro-cosmos, and from which much of man's understanding of the universe has been gleaned. Out of all the objects that he beholds in the universe, the stars are the objects with which he is most familiar, beginning with the most obvious star in the Terran (Earth) sky, the Sun (Sol). The remnants of long-dead exploded stars (supernovae) are what provided the matter

D. Brown (✉)
Specola Vaticana, V-00120, Città Del Vaticano, Holy See (Vatican City State)
e-mail: dbrown@specola.va



G. Gionti, S.J. and J.-B. Kikwaya Eluo, S.J. (eds.), *The Vatican Observatory, Castel Gandolfo: 80th Anniversary Celebration*, Astrophysics and Space Science Proceedings 51, https://doi.org/10.1007/978-3-319-67205-2_5

from which our own solar system, including the Earth and Sun, formed. It is by the energy derived from the Sun that life on this planet was and is given a stable environment in which to arise and to thrive. Ironically, it is also by the Sun's own evolution that our planet Earth will one day come to its end about five billion years from now. One hopes that humanity will have found a new home planet by then.

Given that the stars are the most prominent and visibly numerous objects observed in the night-time sky, it is no surprise that much of man's understanding (the subject of astrophysics, which seeks to understand the nature of astronomical objects) of the cosmos and what it contains has been derived by studying in great detail the light from stars that arrives here on Earth as collected by a vast array of telescopes. One of the most important advances in stellar astrophysics was brought about by Fr. Pietro Angelo Secchi, S.J. (1818–1878), a Jesuit astronomer (see Fig. 1) working at the Pontifical Observatory of the Roman College, the predecessor to the Vatican Observatory. Applying the methods of spectroscopy to astronomy, he devised the first system of how to classify stars according to the spectral profile of their emitted light as seen in Fig. 2. This inaugurated a new era in how to understand the chemical composition of stars as well as understanding something of their evolution, at least in

**Fig. 1** Fr. Angelo Secchi, S. J. (1818–1878)

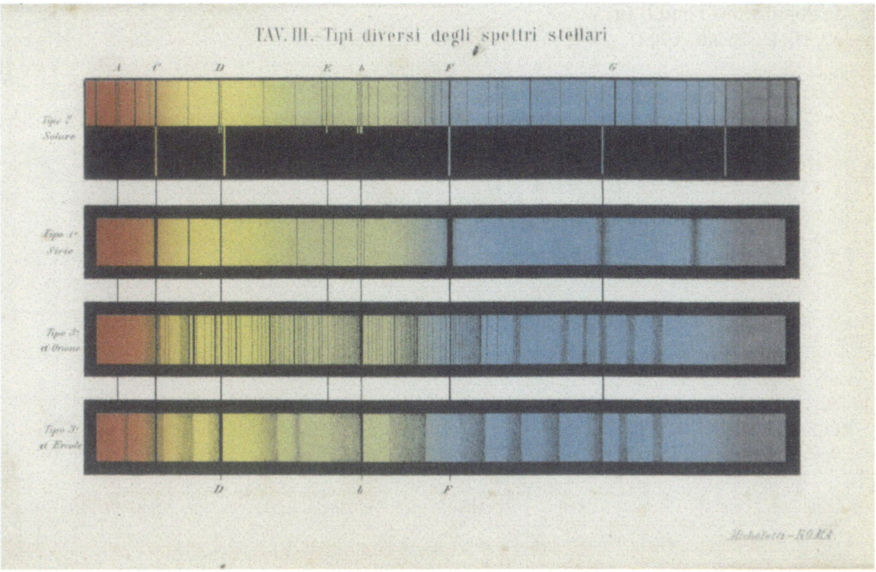


**Fig. 2** A plate of spectra from Secchi's book, *Le Stelle*, by which stars can be classified (Secchi 1877)

some aspects for that era. This, in turn, was instrumental in how to understand large groupings of stars, known as stellar populations.

## Introduction to Stellar Populations

A stellar population is understood to be a large collection (group) of gravitationally-bound stars and gas which share common properties of age, chemical abundance, location, and kinematics (motions). The concept of such a structure arose from observations that galaxies, such as the Milky Way Galaxy, consist of component collections of stars and gas in which the objects in each grouping share the properties mentioned above. In 1944 such observations led to a formalized scheme by Walter Baade. By observing galaxies such as M31, M32, M110 (NGC 205), and the Milky Way, he was able to identify two distinct and general types of stellar populations (Population I and Population II) within them on the basis of how their component stars appear on HRD/CMD (Hertzsprung-Russell Diagram/Color Magnitude Diagram) diagrams (Baade 1944) as seen in Fig. 3. Another seminal moment in the development of the idea of stellar populations came in 1957 with the Vatican Conference on Stellar Populations: Semaine d'Etude sur Le Problème des Populations Stellaire. It was sponsored by the Pontifical Academy of Sciences during May 20–28, 1957, and it was held in order to study the problem of stellar populations and galactic structure, calling together some of the greatest astronomers of that time,

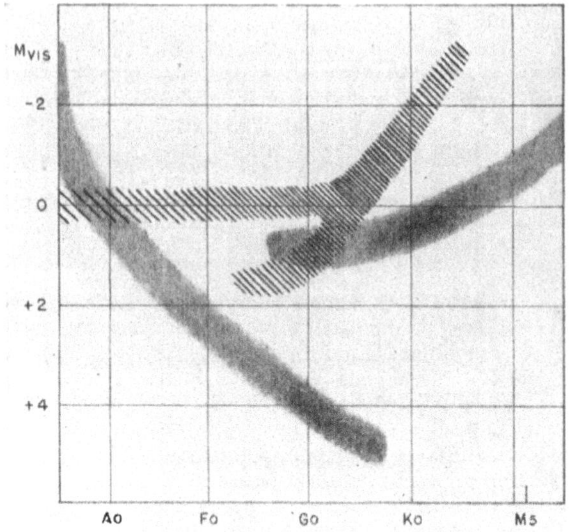


**Fig. 3** Populations I and II in Baade's HRD (Baade 1944)

including Baade, Blaauw, Hoyle, LeMaître, Morgan, Oort, Salpeter, Sandage, Schwarzschild, Spitzer, Strömgren, etc. (Fig. 4). The group was received by Pope Pius XII on May 20, 1957, and the proceedings from the conference were published shortly thereafter in the volume Stellar Populations under the auspices of the Pontifical Academy of Sciences and edited by the director of the Vatican Observatory at that time, Fr. D.J.K O'Connell, S.J. (O'Connell 1958).

**Fig. 4** Participants in the 1957 Vatican Conference on Stellar Populations (O'Connell 1958)

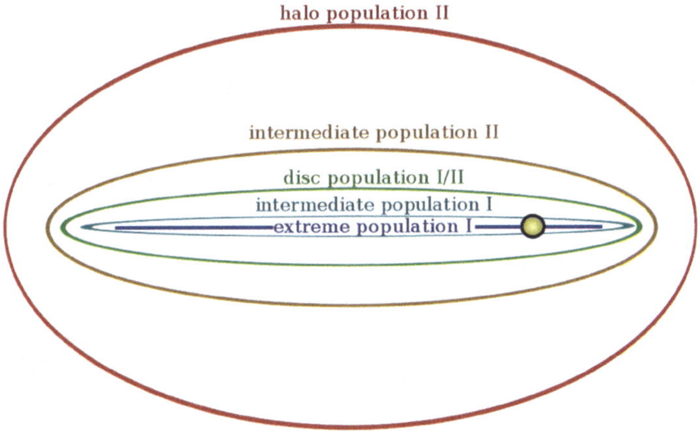


**Fig. 5** Distribution of stellar populations in the Milky Way Galaxy according to the 1957 Vatican Conference (Larsson-Leander 1970)

At that time, the results of the conference represented an advance in the understanding of stellar populations. In particular, the conference provided a systematic and coherent way to understand a stellar population according to age (decreasing age), kinematics (increasing flattening), and chemical composition (increasing metal abundance) of component stars. Moreover, according to such criteria, it identified and defined five basic Galactic populations into which stars are distributed in the Milky Way Galaxy: the halo (Pop. II), an intermediate Population II grouping, a disk population (Populations I and II), an intermediate Population I grouping, and an extreme Population I grouping (see Fig. 5).

Since the mid-twentieth century, the concept of a stellar population, in regard to Galactic structure (especially for spiral galaxies [resembling hurricanes with spirals bands] in which the distinctions are clearer than in elliptical [watermelon-shaped, ellipsoid-like structures]), has undergone further modification, though the basic scheme is still the same (see Fig. 6). In the Milky Way Galaxy, three or four basic subpopulations may be distinguished, depending on how one differentiates them. First, there is the Galactic Halo which is the spheroid distribution of old metal-poor stars and globular clusters and dark matter which extends beyond the disk of the galaxy; there is very little gas or dust or star formation evident in it; component stars have high relative velocities in highly elliptical orbits which take them outside of the Galactic plane. Second, there is the roughly-spherical Galactic bulge, or nucleus, of the Galaxy in which is found a very high density of old Population II stars, gas, and dust; the bulge contains the highest concentration of stars in such a galaxy, and very often it might play host to a super-massive black hole at its very center. Third, there is the Galactic disk which (in spiral galaxies) consists of the planar, or flattened portion, of the galaxy in which are found the spiral arms of the galaxy; it contains hot young Population I stars (with some in open clusters) as well as most of the gas and dust in the Galaxy; it is the setting for star formation. Some

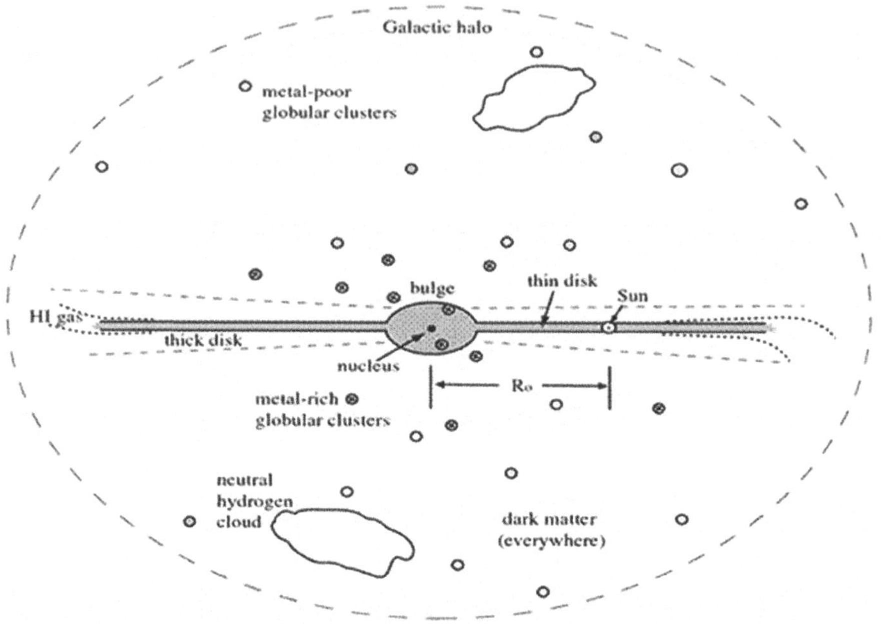


**Fig. 6** The current conception of how stellar populations are distributed within the Milky Way Galaxy (https://www.gaia-eso.eu/science/thick-disc)

schemes of classification distinguish between the thin disk and thick disk. For the Milky Way Galaxy, the Galactic disk has a thickness of about 1000 light-years, a diameter of 100,000 light-years, and makes one complete revolution in about 250,000,000 years.

Among the stellar populations associated with galaxies (which are composite populations of subpopulations), their either being found within them or gravitationally bound to them in one way or another, open clusters within the galactic disk and globular clusters in the galactic halo, respectively, often are of particular interest in stellar evolution. Such clusters reveal much about the evolutionary history of the stars that constitute them. Of considerable help in this is the use of the HRD (or the CMD as the observational equivalent using photometric colors) of a given cluster, which in effect shows the evolutionary state (MS: main sequence, RGB: red giant branch, HB: horizontal branch, AGB: asymptotic giant branch, WD: white dwarf) each star in a cluster for a given time (Fig. 7). This, of course, depends on seeing open and globular clusters as simple stellar populations (SSPs) in which the stars in such clusters can be taken to have been formed at roughly the same time in an initial single burst of star formation. The issue of star formation complicates the HRD diagrams for composite stellar populations such as galaxies because of the different ages of their component populations. Do, note, however, that considering a globular cluster a simple stellar population (SSP) is an oversimplification. Recent observations indicate the existence of multiple

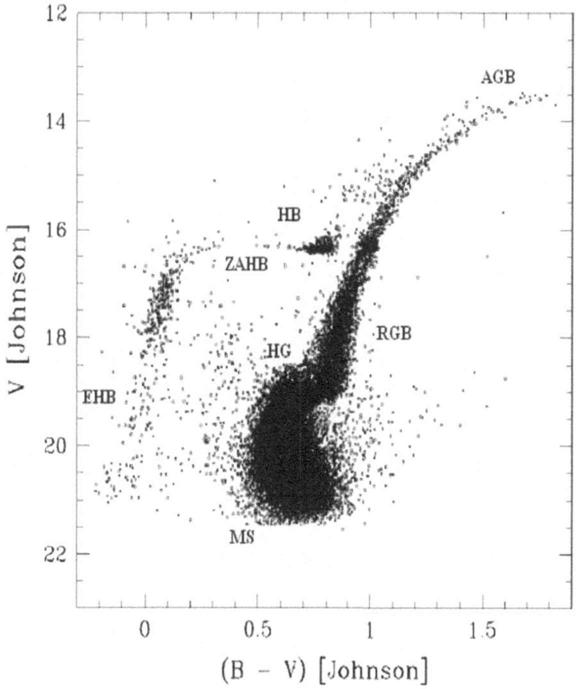


**Fig. 7** A prototypical HRD/CMD globular for a globular cluster (NGC 2808) that shows stars in different evolutionary phases: MS (main sequence), HG (Hertzsprung gap), RGB (red giant branch), HB (horizontal branch), ZAHB (zero-age horizontal branch), EHB (extreme horizontal branch), AGB (asymptotic giant branch) (Sosin et al. 1997)

helium-enriched subpopulations in such structures, which implies multiple episodes of star formation, one example of this being Omega Centauri.

## Some Ongoing Research in Stellar Populations at the Specola Vaticana

One important area of ongoing research in the area of stellar evolution and stellar populations at the Vatican Observatory concerns the study of hot subdwarf stars. These are hot blue stars that constitute the dominant population of faint blue stars at high Galactic latitudes and in giant elliptical galaxies. They can be identified in the HRD as located between the upper MS and the WD (white dwarf) sequence, and they can be differentiated into three types according to the spectral features of H and He on their surfaces: sdB, sdO, and sdOB stars, with temperatures in the range of 23000 K $\lesssim T_{eff} \lesssim$ 35000 K (Heber et al. 1984; Heber 1986; Moehler et al. 1990; Theissen et al. 1993; Edelmann et al. 2003), 30000 K $\lesssim T_{eff} \lesssim$ 40,000 K (Baschek and Norris 1975; Groth et al. 1985; Heber 1986), and 40000 K $\lesssim T_{eff} \lesssim$ 80,000 K, respectively.

Returning once again to the HRD in Fig. 7 and also using the luminosity-independent ($\log_{10}g$, $\log_{10}T_{eff}$) diagram (which graphs gravity $\log_{10}$ g versus temperature $\log_{10}T_{eff}$, as opposed to the HRD which graphs luminosity $\log_{10}L$ versus temperature $\log_{10}T_{eff}$), as done by Newell (1973), sdB stars in the Galactic field (stars not belonging to clusters) can be identified with and occupy the same region on the diagram as the hottest blue stars on the horizontal branch (HB) of globular clusters; that is, they can be identified with extreme horizontal branch (EHB) stars. As such, from an astrophysical point of view, hot subdwarf stars are core He-burning (H-burning = nuclear fusion of hydrogen in the stellar core [center] when stars are young; He-burning = nuclear fusion of helium in the stellar core when stars have exhausted H and are 'middle-aged') stars with a core mass of $\sim 0.5\ M_{\odot}$ ($M_{\odot}$ = mass of the Sun) and very thin hydrogen envelopes (which surround the inner stellar core) with $M_{env} \lesssim 0.02\ M_{\odot}$. Such thin envelopes imply that EHB/sdB stars will not evolve as most normal stars do by continuing to the asymptotic giant branch (AGB) phase (see Fig. 7) of evolution in the post-He-burning (in their cores) phase of their evolution. Instead, they will evolve directly into being white dwarf (WD) stars (no more nuclear fusion in the stellar core) on the WD cooling curve, the locus of points on the HRD where WD stars are located.

The thin hydrogen envelopes of hot subdwarf stars will also be subject to various diffusive processes within them, processes by which the original element composition of the envelope is changed because of the transport and redistribution of atoms of different elements due to different factors such as photon-atom interactions (radiative levitation) and gravity (gravitational settling). These can lead to unique surface chemical abundances in stellar atmospheres/envelopes and also to pulsations in some hot subdwarf stars (Kilkenny et al. 1997; Charpinet et al. 1996). Asteroseismological (the study of pulsations) studies of these stars, in turn, can provide information on several properties of sdB stars, including interior stellar structure and stellar mass, quantities that are not easy to determine in many stars. The latter can be used to yield a preliminary empirical mass distribution (how the masses vary statistically) of such stars that can then be compared to the theoretical mass distribution predicted by models in order to ascertain the different ways by which hot subdwarf stars form and evolve (formation channels). Recent studies indicate that the empirical mass distribution of Fontaine et al. (2012), which shows a high number of stars at around $M \sim 0.47\ M_{\odot}$, agrees well with that predicted by Han et al. (2002, 2003), which has most subdwarfs having masses in the range $M \sim 0.44\ M_{\odot}$–$0.50\ M_{\odot}$.

As mentioned previously, stars in particular groupings can retain at least some of the properties of the stellar populations from which they originated in terms of their distinct chemical (e.g. metallicity Z, which is the mass fraction of a star that is not hydrogen and helium) signatures, or kinematic (orbital velocities, velocity dispersion [variation in velocities about the mean velocity]) motions, or spatial distribution, or age properties. Hot subdwarf stars are found in all Galactic populations (halo, bulge, thick/thin disc), including in globular clusters, all four of which have different chemical histories (Altmann et al. 2004; Napiwotzki et al. 2004) and kinematics. Moreover, they are also found in other galaxies, including in giant elliptical galaxies. In this extragalactic context, it is thought that hot subdwarf stars themselves give rise to the ultraviolet upturn (UV-upturn or UVX) phenomenon observed in many giant elliptical galaxies, bulges of spiral galaxies, and even in some clusters of stars (Fig. 8). Basically, it is a spike of ultraviolet energy in the light spectrum of such a galaxy—unexpected because such old galaxies have few stars to produce light of those frequencies.

Given that hot subdwarf stars are thought to be the field equivalent of EHB stars in globular clusters, they are thought to be the result of progenitor stars which lost most of their outer hydrogen-rich envelopes during the red giant branch (RGB: outer envelope of the star expands after H fusion stops in central core) stage of evolution just prior to the rapid onset of He fusion their cores (the helium flash). The thinner the surrounding hydrogen envelope of the resulting star, the hotter it is on the horizontal branch (Fig. 7) as one moves to the left in the HRD, especially EHB stars. However, the physical mechanism—thus the evolutionary channel—by which such extreme envelope mass loss occurs remains unknown and an item of contention. Does stellar environment play a role, and, if so, how does one account for the fact that such sdB stars are found in diverse stellar populations, for example both in the galactic field and in globular clusters? Both their comparison to EHB stars and also the suspicion that they might also give rise to the UVX in elliptical galaxies have over time given rise to two basic competing scenarios by which sdBs form: single star formation channels and binary star formation channels. On the one hand, unique physical processes in single stars might be responsible for causing extreme mass loss from the surface of the progenitor star which will produce an EHB/sdB star in one of four ways: (1) an EHB star results from a RGB star from which emanates a very high stellar wind, the rate of mass loss being described theoretically by the Reimers wind loss formula with a fairly broad distribution in

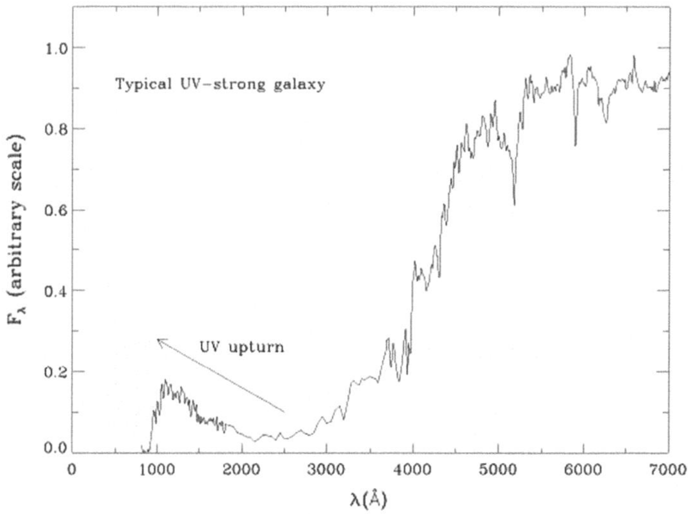


**Fig. 8** The UV upturn phenomenon in the composite spectrum of a UV-strong galaxy, NGC 4552 (Yi et al. 1998)

the η-mass loss parameter for different metallicities (D'Cruz et al. 1996); (2) flash mixing whereby a star experiences a delayed He flash after extreme mass loss (Sweigart 1997); (3) a HB star undergoes mass loss during the HB phase itself (Yong et al. 2000); (4) in globular clusters such as ω-Centauri, a Na-O anti-correlation and multiple main sequences suggest a fraction of helium-enriched stars (subpopulations) that formed from the AGB ejecta of primordial stars in the cluster, thus producing very hot EHB stars (Fig. 9 from Lee et al. (2005)) as suggested by D'Antona and Caloi (2008) and Piotto et al. (2007).

On the other hand, various binary star scenarios have also been proposed by which a hot subdwarf star results from the interaction of two stars in a binary system, in particular when one star (the sdB progenitor) in the binary system overflows (as it expands during its evolution) its gravitational sphere of influence (Roche lobe: inside of which its material is gravitationally bound to the star) and transfers mass (RLOF: Roche lobe overflow) to its companion star (Mengel et al. 1976). This hypothesis is compelling and attractive, given that approximately 50% of stars are thought to be in binary systems and that binary interactions provide a naturally physical explanation. Moreover, observations indicate that 2/3 of EHB/sdB stars in the Galactic field are found in short-period binary systems (Maxted et al. 2001).

Among the proposed binary formation channels, three common evolutionary channels for sdB stars were explored by Han et al. (2002, 2003): (1) a resulting sdB star in a short-period binary system formed from ejection of a common envelope (CE: cloud of gas that envelopes both stars in a binary when mass transfer from one star to another is too great and unstable to be accreted all at once) resulting from unstable mass transfer from the sdB progenitor to the companion star; (2) a resulting longer-period sdB star resulting from stable mass transfer from the sdB progenitor to the companion; or, (3) from the merger of two helium white dwarfs which result from two episodes of common envelope ejection. Originally, exploration of sdB

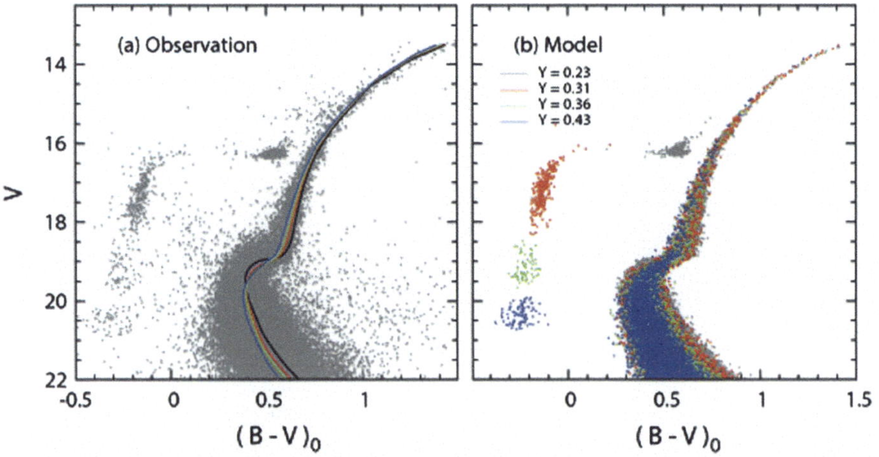


**Fig. 9** A comparison of theoretical models with the observed CMD for distinct subpopulations with different helium abundances in NGC2808 (Lee et al. 2005)

formation/evolutionary channels by Han et al. (2002, 2003) was done for stars having a solar metallicity (fraction of elements by mass in a star which is not hydrogen and helium) Z = 0.02. Ongoing research at the Vatican Observatory by Brown extends this work done on binary formation channels to a greater variety of metallicities, principally for Z = 0.0001, 0.001, 0.02, 0.03, 0.04, 0.05, given that sdB stars (including in binaries) exist in different stellar environments. The extent to which chemical composition, such as metallicity or alpha-enhancement, plays a role in the formation of hot subdwarf stars remains unclear (Altmann et al. 2001). On the one hand, some single-star evolutionary models seeking to explain the UVX suggest that chemical environment is a strong factor in the formation of hot dwarfs, given the strength of the UVX in metal-rich populations (Burstein et al. 1988), mainly because of suggested high-mass loss rates in progenitor stars due to metallicity-dependent stellar wind mass loss, which causes the loss of most of the surrounding envelope of the sdB/EHB star before the ignition of helium fusion in its core. One must also consider that the fraction of sdB stars in binary systems in globular clusters, such as NGC 2808, is low ($f \lesssim 0.04$–$0.14$) (Moni Bidin et al. 2011) compared to the fraction ($f \sim 2/3$) of sdBs in binaries in the higher-metallicity Galactic field. Together with recent observations of globular clusters in which sdB stars may be found in possibly multiple helium-enriched subpopulations in such globular clusters (Piotto et al. 2007; D'Antona and Caloi 2008; Lee et al. 2005), the effect of chemical environment cannot be ignored. On the other hand, given the ubiquity of hot subdwarf stars in different stellar populations, and given the dynamics of interactions among the stars in a binary, it seems that, to first order, the metallicity (variation) of stars should not significantly affect the interactions between stars in a binary system, thus the production of sdBs; notwithstanding, metallicity could have an indirect effect via parameters which are metallicity-dependent: (1) stellar wind mass loss that varies with the metallicity Z of a star; (2) stellar evolutionary rate of a star varies with Z; (3) different core masses of stars close to the end of the RGB phase of evolution varying with Z; and, (4) different opacities (the degree of transparency of the gases to the transit light/photons) which affect the extent of the envelopes of RGB and HB stars.

Ongoing research by Brown at the Specola suggests that, to first order, metallicity (chemical composition of a star) does not have a dramatic effect on sdB star formation via binary evolution channels. The result is logical given that the gravitational sphere of influence (Roche Lobe configuration/geometry) around each star in a binary system is independent of the metallicities of the stars themselves. However, there are subtle secondary effects. Z does definitely affect the evolutionary rate of sdB progenitors, determining when in time sdB stars from particular channels first appear (see Fig. 10)—lower-Z sdB stars will appear before their equivalent mass high-Z sdB counterparts. Changes in Z will also affect the canonical core masses of sdBs and of their progenitor stars, stars with lower Z having slightly higher helium core masses. Z can affect the opacity of the envelopes of HB stars in a greater way, at least before diffusion effects become dominant in EHB stars. Increasing Z increases the opacity of the thin envelope of sdB stars, and, in doing so, lowers the surface temperature of such stars. Higher-Z progenitors in binaries can also produce

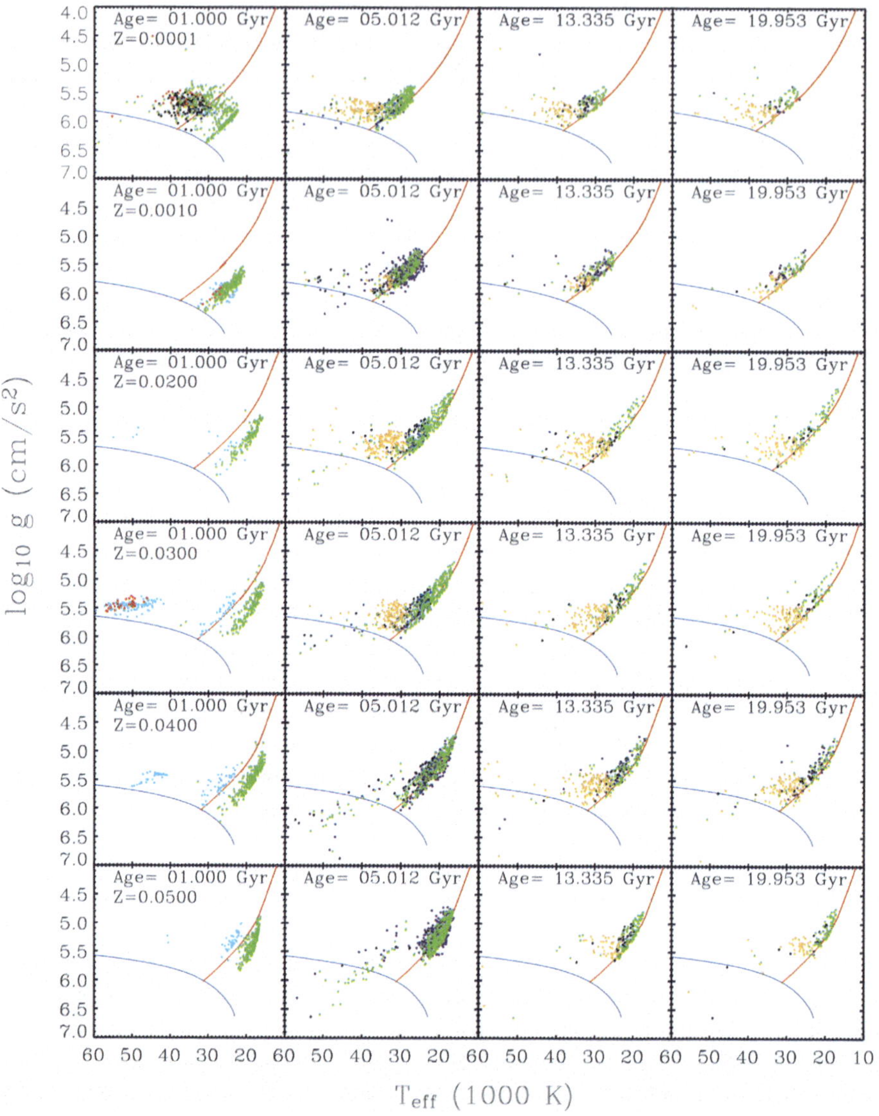


**Fig. 10** sdB time evolution (1–20 Gyr) for simple stellar populations having different Z. Simulated sdB stars from the various formation channels: (1) Stars in black and red are sdB stars which form after one phase of common envelope (CE, unstable mass transfer) formation and ejection; (2) stars dark blue and turquoise result from a second CE phase; (3) stars in green are sdBs that form from a first instance of stable mass transfer in a binary; (4) stars in violet result from a second episode of stable mass transfer in a binary system; (5) sdB stars in yellow-orange are the result of the merger of two WD stars in a binary. The ZAEHB (zero-age extreme horizontal branch) + ZAHB (zero age horizontal branch) are indicated by the red line

longer-period binary sdB stars, mainly because Z, affecting the radius of a sdB progenitor (via the opacity of the envelope) can yield more instances of RLOF in metal-rich stars in long-period binaries at the end of the RGB phase than in a metal-poor systems having the same period but smaller radii. Another effect seen in the variation of Z is the formation of sdB stars from binary systems experiencing a second episode of stable mass transfer (Roche Lobe Overflow) from the donor to accretor star in the case of metal-poor binary progenitors, for which there is no equivalent case for solar metallicity, mainly because such stars would require a WD companion exceeding the Chandrasekhar mass (maximum mass for a WD star beyond which there results a neutron star or black hole).

There are four encouraging results that stem from the binary population synthesis study conducted by Brown done in order to study the production of sdB stars at different metallicities Z. First, as seen in Fig. 11, the theoretical

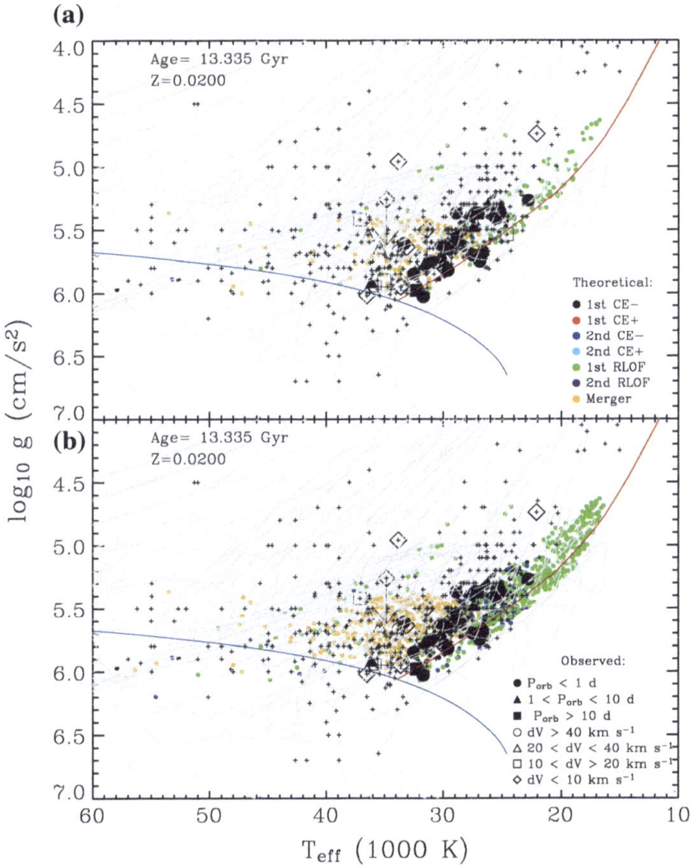


**Fig. 11** A comparison of theoretical sdB stars (in color) with empirically observed sdBs ('+', filled-in and open circles, squares, and triangles, and open diamonds) including data from Saffer et al. (1994), Lisker et al. (2005), and Stroeer et al. (2007)

results cover the same locus in the $\log_{10}g - T_{eff}$-diagram as do the empirical data points. Second, results at different Z provide a possible explanation for sdB stars which are found below the ZAHeMS (Zero-Age Helium Main Sequence; time zero line/locus from which pure helium stars begin to evolve) in a $\log_{10}g$–$T_{eff}$ diagram. They indicate that low-mass sdB stars (0.30 $M_{\odot}$–0.40 $M_{\odot}$) form from intermediate-mass progenitors which have ignited their non-degenerate (non-quantum gas) He-cores as suggested by asteroseismological studies of pulsating stars such as that of Hu et al. (2008) who posit that differences in pulsation modes/frequencies among various pulsating sdB stars stem from differences in internal structure of sdB stars resulting from differences in helium ignition between core degenerate (quantum gas at high density) and non-degenerate sdB progenitor stars. Third, the simulations at different Z are able to model many stellar clusters (such as the open cluster NGC 6791) very well, both in terms of accounting for the production of EHB/sdB stars both in location (which standard models cannot reproduce without ad hoc assumptions) on the HRD and EHB fraction (out of total number of HB stars present). Fourth, the simulations at different Z, when applied to the UVX, support Han et al. (2007)—the UV excess (as measured by the *1550—V* color, measure of UV strength) does not vary much with time (redshift: a measure of distance and also look-back time) in passive (no active star formation) giant elliptical galaxies, at least for low redshifts $z \sim 0$. This is in contrast to single-star models of the UVX, which predict an increase in the UVX for low redshifts.

## Conclusion

The effects of metallicity when studying the formation and evolution of hot subdwarf stars are important, even if Z does not have a dramatic impact on sdB star formation, to first order. The effects are second order insofar as metallicity (1) affects stellar evolutionary rates, and, hence, advent of sdBs from particular channels at different ages, (2) affects the canonical helium core masses and opacities of outer stellar envelopes of their progenitors, (3) yields longer-period sdB binaries in metal-rich scenarios, and (4) they can result in more sdBs in binaries for stable RLOF in metal-poor scenarios where the companion is a non-massive WD. Simulations at different Z also indicate that metallicity does not play a significant role in the evolution of the UVX, though this is by no means definitive. Certainly one area, which promises to yield much information on the evolutionary channels by which sdB stars form, is asteroseismology. In particular by identifying the different oscillations in different parts of pulsating sdB stars, a wealth of information can be learned about such stars, most especially their masses, from which one can then surmise the mass distribution of such stars to compare with theoretical predictions. To this end, future research at the Specola with regard to Brown's work on stellar evolution will move in the direction of using the VATT telescope, together with the GUFI and 4K cameras, in order further to study sdB formation channels. Preliminary observations of the pulsating sdB stars PG 1047+003 and PG 1419+081 have already been done, and the data remain to be processed. This is indeed a promising field of research by which the Vatican Observatory can continue its research in stellar populations.

**Acknowledgements** I thank Drs. Philipp Podsiadlowski, Zhanwen Han, and Sukyoung Yi for lending their expertise generously to various parts of research conducted.

## References

Altmann, M., de Boer, K. S., & Edelmann, H. (2001). The population membership of sdB stars: a kinematic analysis. *Astronomische Nachrichten, 322,* 397.

Altmann, M., Edelmann, H., & de Boer, K. S. (2004). Studying the populations of our Galaxy using the kinematics of sdB stars. *Astronomy & Astrophysics, 414,* 181.

Baade, W. (1944). The resolution of messier 32, NGC 205, and the central region of the andromeda nebula. *The Astrophysical Journal, 100,* 137.

Baschek, B., & Norris, J. (1975). The OB-type subdwarf HD 149382 and the nature of the subdwarf B stars. *The Astrophysical Journal, 199,* 694.

Burstein, D., Bertola, F., Buson, L. M., & Faber, S. M. (1988). The farultraviolet spectra of early-type galaxies. *The Astrophysical Journal, 328,* 440.

Charpinet, S., Fontaine, G., & Brassard, P., Dorman, B. 1996. The potential of asteroseismology for hot, subdwarf B stars: a new class of pulsating stars? *The Astrophysical Journal, 471,* L103.

D'Antona, F., & Caloi, V. 2008. The fraction of second generation stars in globular clusters from the analysis of the horizontal branch. *Monthly Notices of the Royal Astronomical Society, 390,* 693.

D'Cruz, N. L., Dorman, B., Rood, R. T., & O'Connell, R. W. (1996). The origin of extreme horizontal branch stars. *The Astrophysical Journal, 466,* 359.

Edelmann, H., Heber, U., Hagen, H.-J., Lemke, M., Dreizler, S., Napiwotzki, R., et al. (2003). Spectral analysis of sdB stars from the Hamburg Quasar Survey. *Astronomy & Astrophysics, 400,* 939.

Fontaine, G., Brassard, P., Charpinet, S., Green, E. M., Randall, S. K., & Van Grootel, V. (2012). A preliminary look at the empirical mass distribution of hot B subdwarf stars. *Astronomy & Astrophysics, 539,* A12.

Han, Z., Podsiadlowski, P., Maxted, P. F. L., Marsh, T. R., & Ivanova, N. (2002). The origin of subdwarf B stars – I. The formation channels. *Monthly Notices of the Royal Astronomical Society, 336,* 449.

Han, Z., Podsiadlowski, P., Maxted, P. F. L., & Marsh, T. R. (2003). The origin of subdwarf B stars – II. *Monthly Notices of the Royal Astronomical Society, 341,* 669.

Han, Z., Podsiadlowski, P., & Lynas-Gray, A. E. (2007). A binary model for the UV-upturn of elliptical galaxies. *Monthly Notices of the Royal Astronomical Society, 380,* 1098.

Groth, H. G., Kudritzki, R. P., & Heber, U. (1985). Photospheric convection zones and evolution of subluminous OB-stars. *Astronomy & Astrophysics, 152,* 107.

Heber, U., Hunger, K., Jonas, G., & Kudritzki, R. P. (1984). The atmosphere of subluminous B stars. *Astronomy & Astrophysics, 130,* 119.

Heber, U. (1986). The atmosphere of subluminous B stars. II - Analysis of 10 helium poor subdwarfs and the birthrate of sdB stars. *Astronomy & Astrophysics, 155,* 33.

Hu, H., Dupret, M.-A., Aerts, C., Nelemans, G., Kawaler, S. D., Miglio, A., et al. (2008). A seismic approach to testing different formation channels of subdwarf B stars. *Astronomy & Astrophysics, 490,* 243.

Kilkenny, D., Koen, C., O'Donoghue, D., & Stobie, R. S. (1997). A new class of rapidly pulsating star - I. EC 14026-2647, the class prototype. *Monthly Notices of the Royal Astronomical Society, 285,* 640.

Larsson-Leander, G. (1970). Astronomi och astrofysik. (https://commons.wikimedia.org/wiki/File:Starpop.svg).

Lee, Y.-W., Joo, S.-J., Han, S.-I., Chung, C., Ree, C.-H., Sohn, Y.-J., et al. (2005). Super-helium-rich populations and the origin of extreme horizontal-branch stars in globular clusters. *The Astrophysical Journal, 621,* L57.

Lisker, T., Heber, U., Napiwotzki, R., Christlieb, N., Han, Z., Homeier, D., et al. (2005). Hot subdwarfs from the ESO Supernova Ia Progenitor Survey. I. Atmospheric parameters and cool companions of sdB stars. *Astronomy & Astrophysics, 430,* 223.

Maxted, P. F. L., Heber, U., Marsh, T. R., & North, R. C. (2001). The binary fraction of extreme horizontal branch stars. *Monthly Notices of the Royal Astronomical Society, 326,* 1391.

Mengel, J. G., Norris, J., & Gross, P. G. (1976). A binary hypothesis for the subdwarf B stars. *The Astrophysical Journal, 204,* 488.

Moehler, S., Heber, U., & de Boer, K. S. (1990). Hot subluminous stars at high galactic latitudes. II - Physical parameters and distances of 37 sdB stars. *Astronomy & Astrophysics, 239,* 265.

Moni, Bidin C., Villanova, S., Piotto, G., & Momany, Y. (2011). A lack of close binaries among hot horizontal branch stars in globular clusters. II. NGC 2808. *Astronomy & Astrophysics, 528,* A127.

Napiwotzki, R., Karl, C. A., Lisker, T., Heber, U., Christlieb, N., Reimers, D., et al. (2004). Close binary EHB stars from SPY. *Astrophysics and Space Science, 291,* 321.

Newell, E. B. (1973). The evolutionary status of the blue halo stars. *The Astrophysical Journal Supplement Series, 26,* 37.

O'Connell, D. J. K. (ed.) (1958). *Stellar Populations: Proceedings of the Conference Sponsored by the Pontifical Academy of Sciences and the Vatican Observatory.*

Piotto, G., Bedin, L. R., Anderson, J., King, I. R., Cassisi, S., Milone, A. P., et al. (2007). A Triple Main Sequence in the Globular Cluster NGC 2808. *The Astrophysical Journal Letters, 661,* L53.

Saffer, R. A., Bergeron, P., Koester, D., & Liebert, J. (1994). Atmospheric parameters of field subdwarf B stars. *The Astrophysical Journal, 432,* 351.

Secchi, A. (1877). Le Stelle: Fratelli Dumolard.

Sosin, C., Dorman, B., Djorgovski, S. G., Piotto, G., Rich, R. M., King, I. R., et al. (1997). Peculiar multimodality on the horizontal branch of the globular cluster NGC 2808. *The Astrophysical Journal, 480,* L35.

Stroeer, A., Heber, U., Lisker, T., Napiwotzki, R., Dreizler, S., Christlieb, N., et al. (2007). Hot subdwarfs from the ESO supernova Ia progenitor survey. II. Atmospheric parameters of subdwarf O stars. *Astronomy & Astrophysics, 462,* 269.

Sweigart, A. V. (1997). Effects of helium mixing on the evolution of globular cluster stars. *The Astrophysical Journal Supplement Series, 474,* L23.

Theissen, A., Moehler, S., Heber, U., & de Boer, K.-S. (1993). Hot subluminous stars at high galactic latitudes – IV. Physical parameters and distances of 18 hot subdwarf stars and their spatial distribution. *Astronomy & Astrophysics, 273,* 524.

Yi, S. K., Demarque, P., & Oemler, A. J. (1998). On the origin of the ultraviolet upturn in elliptical galaxies. II. Test of the horizontal branch hypothesis. *The Astrophysical Journal, 492,* 480.

Yong, H., Demarque, P., & Yi, S. (2000). Mass loss on the horizontal branch: an application to NGC 6791. *The Astrophysical Journal, 539,* 928.

# The Vilnius Photometric System—Studying Stars and Interstellar Matter at the Vatican Observatory

**Richard P. Boyle, S.J. and Robert Janusz, S.J.**

**Abstract** We introduce stellar photometry, its purpose and relationship to other astronomical quantities, presenting it within the context of astronomical research at the Vatican Observatory. We demonstrate the usefulness of the Vilnius Photometric System previously shown at Vilnius Astronomical Obs., Edinburgh Royal Obs., and Università di Roma "La Sapienza", and then adopted by the Vatican Observatory for use with its Vatican Advanced Technology Telescope (VATT) on Mount Graham, AZ, USA. The development of astronomical observations has led from photographic plates, through photoelectric detectors to the use of sensitive CCD cameras. Expanding the photoelectric databank of "Vilnius Stars", we describe the development of new techniques of data-acquisition, calibration to photometric standards, stellar classification and interstellar-matter characterization. Looking ahead, we briefly describe the extensive data set, and our plans for further detailed analyses.

## General Ideas on Stellar Photometry

Astronomical photometry measures incident starlight. For ancient people it was rather simple: they compared the perceived brightness of stars. Having no idea of the astrophysical causes of what they saw, they considered stars to be bright points, fixed to the celestial vault. Wandering celestial bodies (planets) with noticeable proper motion had their own spheres. The Sun was one of them because it could not be described as fixed to the celestial vault.

How to measure light? What is light? Such questions have been asked since ancient times. Applying the scientific method to these questions has led us to a new

Richard P. Boyle, S.J. (✉) · R. Janusz, S.J.
Vatican Observatory (Specola Vaticana), Castel Gandolfo, Italy
e-mail: rpboyle@mac.com

R. Janusz, S.J.
Jesuit University Ignatianum in Kraków, Kraków, Poland



G. Gionti, S.J. and J.-B. Kikwaya Eluo, S.J. (eds.), *The Vatican Observatory, Castel Gandolfo: 80th Anniversary Celebration*, Astrophysics and Space Science Proceedings 51, https://doi.org/10.1007/978-3-319-67205-2_6

understanding of the world and of ourselves. From the diligent study of starlight we have learned that our bodies are made of material coming from evolving stars.

The ancient Greeks measured stellar brightness by eye, placing all visible stars on a scale of *magnitudes* ranging from 1 (the brightest) to 6 (the faintest). With our current sophisticated detection systems we have learned about the relationship between the observed (apparent) stellar magnitude ($m$) and the physical (absolute) magnitude ($M$) of any particular star. It turns out to be a very complex question, because:

$$m = M + (\text{adjustment for distance}) + (\text{adjustment for extinction}).$$

In other words, this apparent, observed magnitude ($m$) equals the star's absolute magnitude ($M$) adjusted for the star's distance and for the extinction of starlight caused by its absorption and scattering in interstellar dust.

Unlike the Ancients who thought that all "fixed" stars were at the same distance from us, we know that stars inhabit a three-dimensional space, and therefore are at various distances from us. This translates into the first correction term, the "adjustment for distance". The second correction term, "adjustment for extinction," expresses starlight's attenuation by interstellar matter (mainly dust).

We need to find a method of measurement, allowing us not only to determine the "$m$", i.e. the directly measurable quantity, but also to obtain all the astrophysical quantities: "adjustment for distance" ($D$) and the total extinction ($A$) of light, i.e. its absorption and scattering by interstellar dust. The problem becomes more difficult, because these astrophysical quantities depend on starlight's color ($\lambda$). This astrophysical approach had a difficult beginning because astronomers could not agree on the new astrophysical methods (see Fig. 1).

An Italian Jesuit, Fr. Angelo Secchi, S.J. (1818–1878), together with other pioneers of astrophysics, introduced the idea of classifying stars on the basis of their spectra. This is the method we use today. His spectra drawings are shown in Fig. 2.

**Popular Astronomy**

Vol. LVIII, No. 4 APRIL, 1950 Whole No. 574

**Fr. Secchi and Stellar Spectra***

By MARTIN F. McCARTHY, S.J.

When the history of the early days of astrophysics, astronomy's newest science, is written, one name which will figure prominently in its pages will be that of Father Angelo Secchi of the Society of Jesus, who for nineteen years directed the Observatory attached to the Roman College. Perhaps, we can anticipate this portion of what should prove a most fascinating story by discussing here the work done by Fr. Secchi in the description and classification of stellar spectra.

**Fig. 1** McCarthy (1950) on the beginnings of astrophysics

A realization of the fact that Fr. Secchi did all of his work visually will doubtless explain why he perceived series of bands in the late-type stars as "kinds of rows of columns." This also increases our admiration for those keen powers of observation and recall which are demonstrated in his drawings of the spectra of different type stars. (Figure 3).

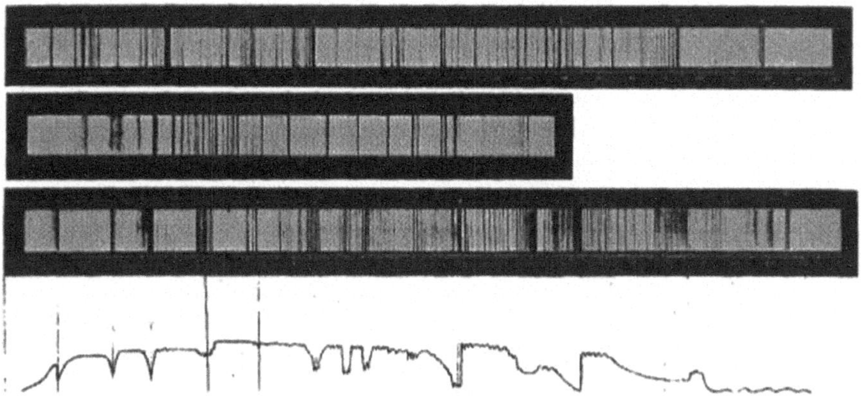

FIGURE 3

Spectrum of α Orionis (Top); α Tauri (Middle); α Scorpii (Bottom). Graph at bottom is Secchi's curve for the variation of light intensity in the spectrum of α Scorpii.

Figures 2 and 3 were photocopied from the prints made of Fr. Secchi's drawings as published in *Sugli Spettri Prismatici delle Stelle Fisi.*

**Fig. 2** Fr. Secchi's spectra (McCarthy 1950)

Although this is an early example, we can see what Secchi started to develop. With a very simple spectroscope he studied "colors" of the stars and their differences. It turns out that the spectral class of a given star plays a role in our equation (above).

Today, with much better digital instruments, we obtain much more accurate digitized spectra from which we derive better stellar classification (Gray and Corbally 2009). It is not easy to determine the spectra of the stars themselves because their light is modified by the effects of distance and extinction due to dust. What is more, to study fainter stars, we must sacrifice the spectral details and measure stellar colors across color bands. These bands need to be chosen carefully. In practice, we use filters, which transmit light in a certain range of wavelengths, and are opaque to light of other wavelengths. The goal is to use a filter set with the minimum number of filters for efficiency, but capturing all the critical spectral regions informative of star types. How to devise a filter set that would work for all types of stars? This was the goal of Lithuanian astronomers from the Vilnius Observatory, which by the way was established by Jesuits in the 18th century.

Moreover, the filter set must allow us to disentangle information about the star and the effects of extinction due to dust. Extinction depends on the wavelength, affecting the blue light more than the red light. The effect is known as interstellar reddening. Our goal is to determine the extinction *A*. Since *A* depends on the star's "color" or spectral class, we must first devise a way of determining the spectral class. To solve this problem we need a minimum of three filters (passing over some theoretical presuppositions not essential for our level of explanation). And indeed, the classical *UBV* filter system has been very successful in determining the intrinsic "color" of a star, regardless the interstellar medium. Because the dust is attenuating the blue part of the spectrum more than the red, the star seems to be redder and this effect is called interstellar reddening. The interstellar reddening-free *Q*-parameter indicates (at least at first glance) the stellar class and the quantity of interstellar dust.

The real problem is in the details: not all stars can be classified with the *UBV* broad band system. To resolve this question, many photoelectric systems were proposed. One of them is the Vilnius seven-color intermediate-band Photometric System. The filter positions with respect to the basic spectral types of stars are shown in Fig. 3.

To continue our reflection on the spectral analysis let us now introduce a quantity reflecting the temperature of a given star. The main "color index" can be understood in terms of black-body radiation, which is a good approximation of some types of stars (Fig. 4).

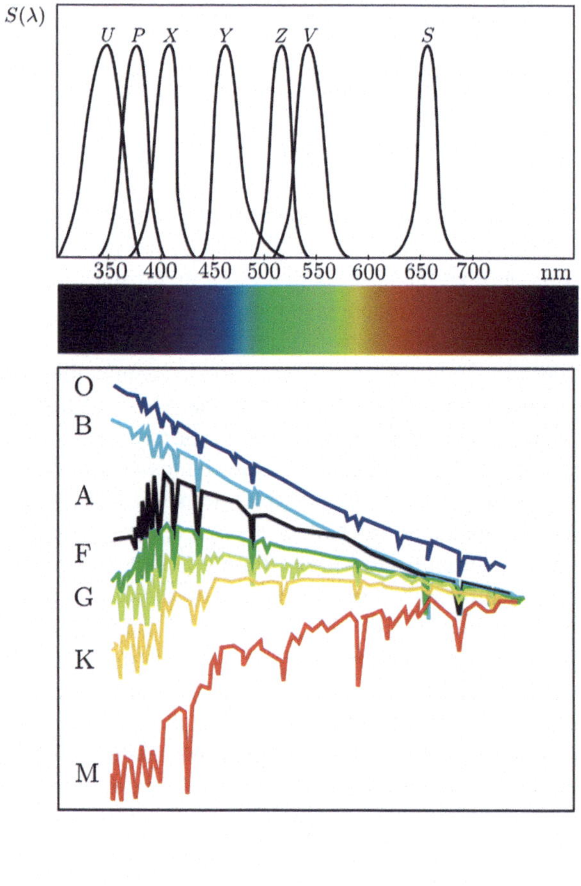


**Fig. 3** The spectra of different star classes (OBAFGKM) and the filter positions corresponding to key spectral features

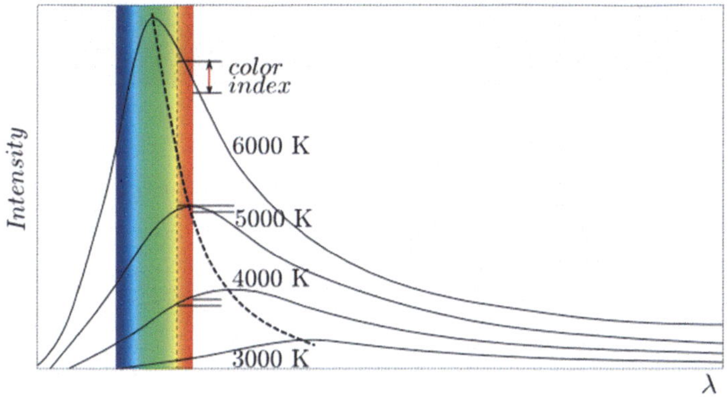


**Fig. 4** Black-body spectra depend on temperature. Notice the variation of the maximum's position

In this case, the difference between end of yellow and end of red color bands, tells us something about the position of maximum on a plot indexed by black-body temperatures. As we can see on this simple example, there are regions of the wavelength which give us better information on the "stellar" temperature, i.e. better to determine the color of the star.

When we add a third filter, in the region where the star light is more absorbed by interstellar dust, we will have the second color index, i.e. a difference between magnitudes given by three filters, which are now "dust" dependent. If we use filters *UYV*, we can construct these indexes as $YV = m_Y - m_V$ and $UY = m_U - m_Y$ which construct a reddening free parameter:

$$Q = UY - \chi_{UYV} YV$$

where $\chi_{UYV}$ depends on filters and is constant for the system (can be calibrated from known standards with well-established spectra).

In Fig. 5 we see how the interstellar matter, interacting with the star light, is attenuating the left part of our spectra more than the right one. The $Q$ parameter measures this effect, but the price is that it depends on at least three filters.

## The Vilnius Photometric System

The ideas described above give us some intuition on how the Vilnius System works. Here is its graphical presentation comparing it to the broad-band *UBV* (Fig. 6). Below we show why a single $Q$ parameter is not sufficient to cover all star types (Fig. 7).

Let us consider the A0V standard star Vega. Its spectrum gives all possible information for such a bright nearby star. The broad-band filters (Fig. 6), because of

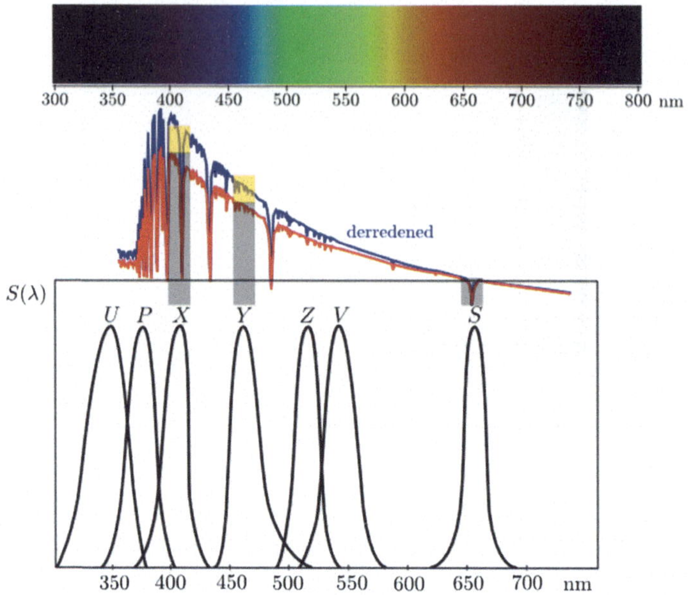


**Fig. 5** The idea of the $Q$-parameter to measure the squashing of the spectra

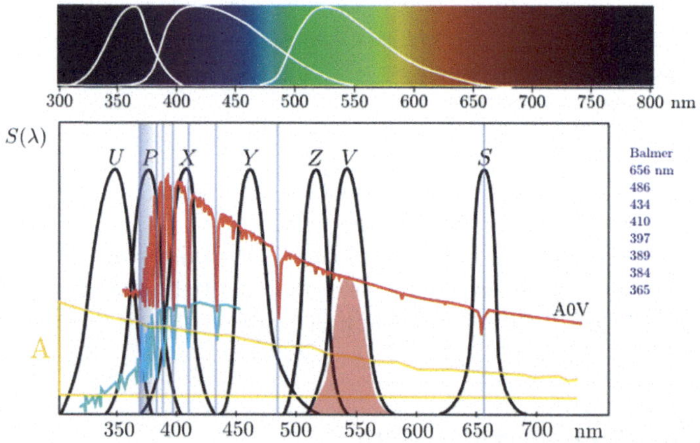


**Fig. 6** UBV filters overlaying the Vega-type star spectrum with Vilnius filters *UPXYZVS* and extinction law *A*; the Balmer jump is around the *P* filter

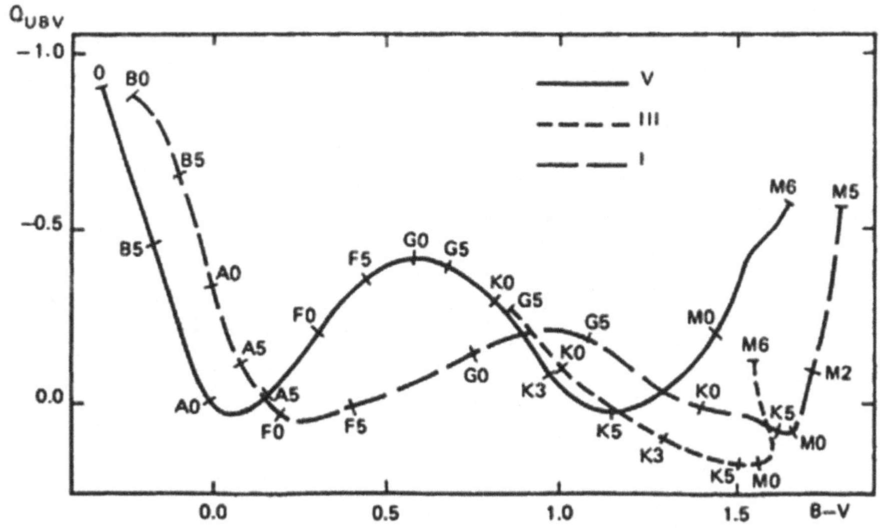


**Fig. 7** The *UBV* system with its *Q* parameter cannot classify all spectral types unambiguously (is not a function): V, III and I are the luminosity classes (Straižys 1992)

the width of their pass band, fail to respect and reveal various key details of the spectrum. On the other hand, the intermediate-band Vilnius filters by design capture the feature-rich spectral bands, for example the H alpha in the *S* filter and the Balmer jump in the *P* filter. The extinction law is shown in yellow color (Fig. 6).

The three-filter *UBV* system yielding only one *Q* parameter cannot classify all spectral types of stars, as shown in Fig. 7. The seven-filter Vilnius system with five basic reddening-free *Q* parameters provides stellar classification of all spectral types.

With multiple *Q* parameters, we can plot not only one *Q* color diagram but many *Q*-*Q* function diagrams allowing us to determine spectral and luminosity classes of stars better (Fig. 8).

The Vilnius system started to be formed around 1963 (Straižys 1963). Successive years were dedicated to its important improvements in calibration (Bull. Vilnius Obs. 1970; Straižys 1970, 1973). This excellent System, created by Vytautas Straižys and the Vilnius group, had unfortunately its vicissitudes related to various factors (geopolitical, distrust, etc.), but we simply skip them. We focus here only on some scientific ideas of the system, important for its implementation at Specola. The usefulness of the system was shown on many ways by V. Straižys and his group at the Vilnius Astronomical Obs., by K. Nandy and his collaborators at Edinburgh Royal Obs., and F. Smriglio with his group at the Università di Roma "La Sapienza". Later on, other institutions were interested in the system: the United States Naval Observatory Flagstaff Station—F. Vrba, Union College—A.G. Davis Philip, Complejo Astronómico El Leoncito (Casleo)—O. Pintado and others.

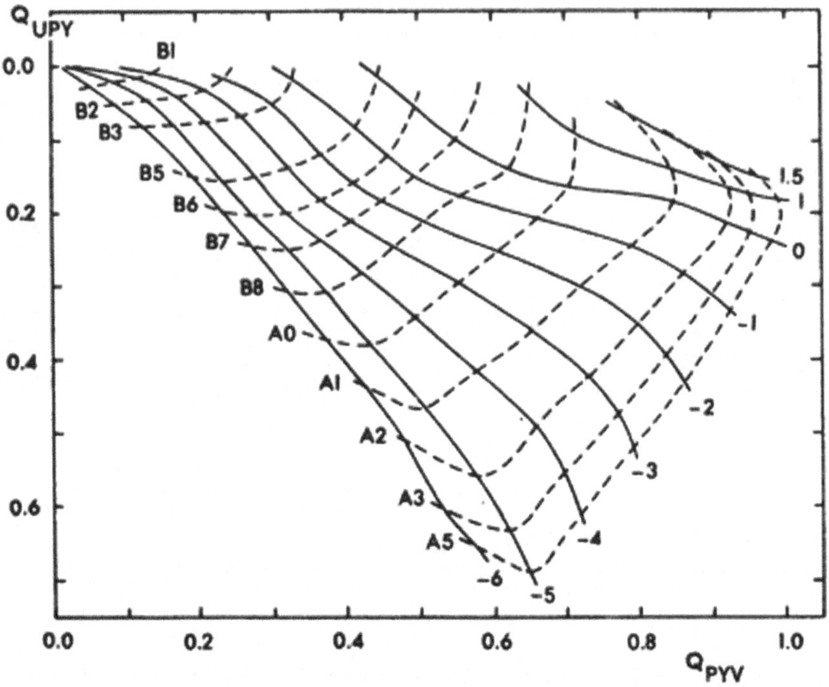


Fig. 84. The $Q_{UPY}$, $Q_{PYV}$ diagram calibrated in spectral classes and absolute magnitudes $M_V$.

**Fig. 8** Vilnius System *Q*-*Q* parameters reveal the spectral classes and the absolute magnitude (Straižys 1992)

## The Vilnius System at the Vatican Observatory

McCarthy et al. (1977) have seen the problem of new photometry. In the context of the IAU Commission 25 "Stellar Photometry and Polarimetry," the topic was discussed by M.F. McCarthy, V. Straižys and G.V. Coyne in 1979 (McCarthy 1979). Fr. McCarthy, S.J., started then to think about the criteria for spectral classification, but no serious project started. Then K. Nandy and F. Smriglio suggested the Vilnius System to G.V. Coyne, S.J.—the Director of the Specola. After his assignment to the Vatican Observatory in 1981, R.P. Boyle, S.J. started to work with the already existing group (Straižys, Nandy, Smriglio) with Vilnius System not only using the Schmidt photographic plates but later on also new methodology of CCD cameras. The Vatican Advanced Technology Telescope (VATT) just from its beginnings (ca. 1996/8) was capable to adapt this methodology on the basis of previous successful results. However also at Specola, there was some distrust towards the System (see: Corbally and Boyle 1987, and the defense: Smriglio et al. 1990, 1994, 1995). Let us give a cursory glance at our history.

In addition to the headquarters in Castel Gandolfo, the Vatican Observatory established its second site in Tucson, Arizona, USA, in the 1980s. The Vatican Observatory Research Group (VORG) has worked in close collaboration with Steward Observatory (University of Arizona), one of the most important sites for astronomy in the world. Steward's modern instruments became available to papal astronomers. However, the best instrument of the Vatican Observatory was the Schmidt telescope in the papal gardens. A similar Schmidt telescope was used at Campo Imperatore by the Università di Roma "La Sapienza", where the Vilnius System began to be used by Filippo Smriglio. So started the close collaboration between F. Smriglio, K. Nandy, V. Straižys and R. Boyle and their corresponding groups in Italy, Great Britain, Lithuania and USA.

In years 1981–1983 both Schmidt telescopes in Campo Imperatore and Castel Gandolfo observed the M 56 sharing the Vilnius filters to collect together one run. R. Boyle, who did his Jesuit Tertianship in Ireland (1983/4), knew the Royal Observatory in Edinburgh, with its COSMOS digitizer for the photographic plates. The first paper with this photometric system was published in 1986 (Smriglio et al.) and the classification system was set up in Rome.

We reconstructed this study in a simplified way with our methods of analysis, figuring out the transformation equations and classification, using some plates of both Schmidt telescopes (we need seven filters for exact classification) and we arrived to similar results. Plates used for reconstruction (scanned at Specola by A. Omizzolo) were: "V GG14" (Vatican, P2255), "Y Vilnius" ("Campo Imperatore", 1967), "U Vilnius" ("Campo Imperatore", 1972), "S Vilnius" (Vatican, P2317), "Z Vilnius" (Vatican, P2363), "Xa Vilnius" ("Campo Imperatore", 1990), "P Vilnius" ("Campo Imperatore", 1971). Table 1 contains our results.

Photographic plates are now history. Let us examine them. In Fig. 9 we have a fragment of a digitalized Schmidt plate, photographed in Castel Gandolfo.

On the profiles of a star of this plate (Figs. 10 and 12) we can see how rough photographic plates are. Three profiles are shown—number of counts is ca. 4000–5000, and the radial plots of stars are "strange," diverging from a Gaussian profile. The serious problem of photographic plates are the variations of sensitivity on the image surface.

A similar plot (Fig. 11) shows data acquired at Campo Imperatore. The conditions were much better but this is very far from the image quality available today (Fig. 12).

**Table 1** Comparison of star classification between (Smriglio et al. 1986), our analysis and SIMBAD database

| Smriglio et al. (1986) | | | Our result: Sp. & Lum. | SIMBAD (sp_type) |
|---|---|---|---|---|
| RA 1950 | DEC1950 | Sp. & Lum. | | |
| 19 14 12.2 | +30 15 25 | F9 d[warf] | g2 III | G2IV-V |
| 19 14 14.4 | +30 13 31 | K2 d[warf] | k0 IV | K1.5 V |
| 19 14 44.8 | +30 22 53 | F4 d[warf] | f0 IV | F2 V |
| 19 15 20.1 | +30 28 13 | K0 d[warf] | k2 V | K2 V |

**Fig. 9** A photographic image, Castel Gandolfo, Schmidt telescope

The Vilnius Group also used photoelectric instrumentation for the stellar photometry. Among his many successes, Straižys et al. (1996a) combined the Vilnius System with Stromgren, and so the Stromvil system was defined. Straižys got for it in 2000 an AAS Chretien grant, which was also supported by the Vatican Observatory (Philip et al. 2003). The Vatican Observatory with Steward Observatory welcomed the Vilnius specialists to the Mt. Lemmon 1.5-m telescope over a five-year period 2000–2005 for stellar observations with their portable photoelectric photometer for setting up standard stars in this Stromvil System.

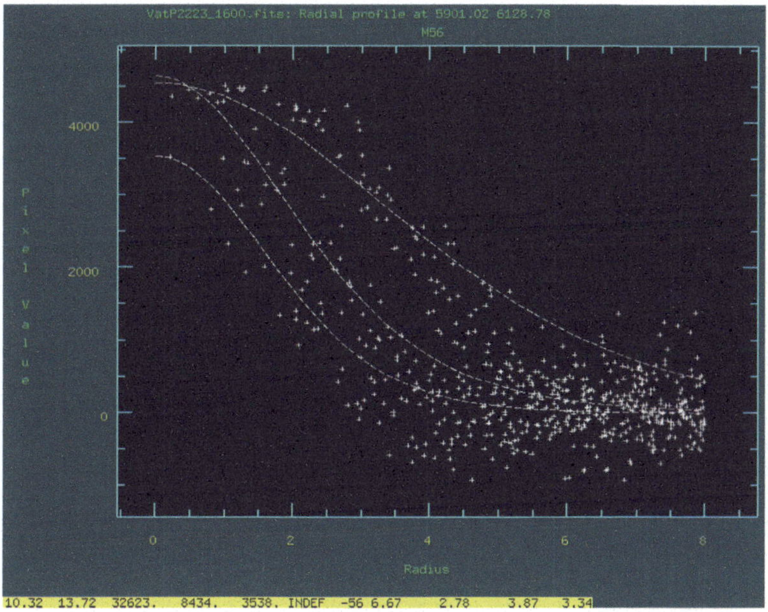

Fig. 10 Non-Gaussian profiles of stars on photographic plates (Castel Gandolfo)

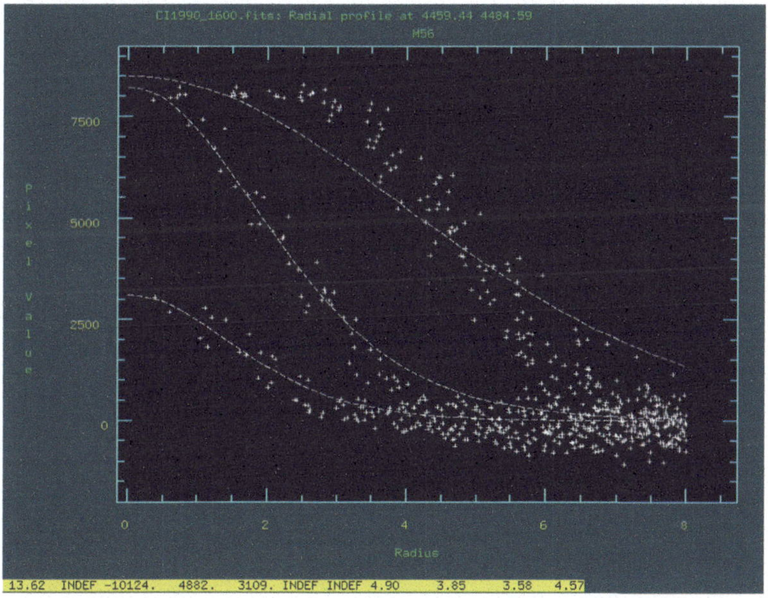

Fig. 11 Non-Gaussian profiles of stars on photographic plates (Campo Imperatore)

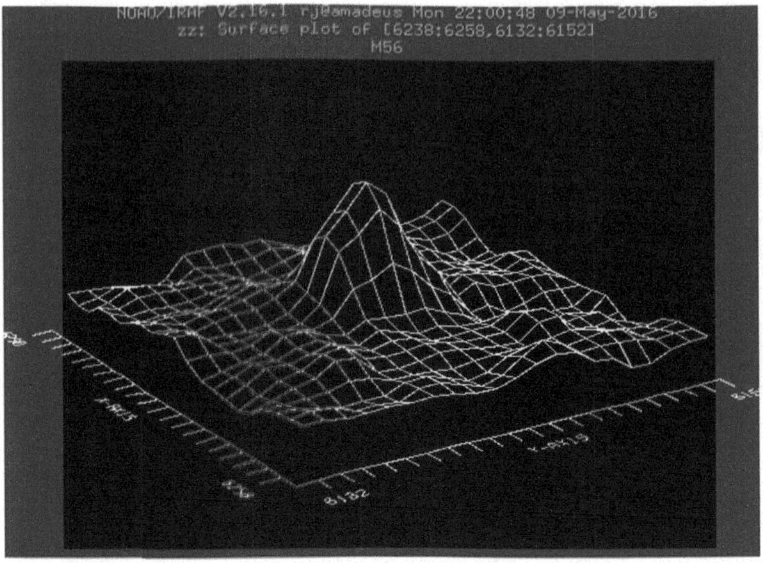


**Fig. 12** A surface profile of a star on a photographic plate

At the same time the Vatican Observatory (with suggestion of Nandy and Straižys) was mainly interested in the very new methodology of the CCD detector available already at Steward Observatory and the US Kitt Peak National Observatory, where R.P. Boyle did it twice (Boyle et al. 1995). The first paper reporting the CCD Vilnius photometry was presented by Boyle et al. in 1990. The observations were performed at Kitt Peak National Observatory in 1986, and the CCD image data processing on KPNO computers. Because the KPNO stopped the continuation of the project (with an excuse, it was not sufficiently tested etc.), Smriglio got access to the telescope of the Università di Bologna (Loiano) till 2006. And also USNO Flagstaff project began. So started a new era for Vatican Observatory photometry.

The success of the Steward Observatory Mirror Laboratory, with its pioneered mirror technology, opened new opportunities for the project of a new Vatican Telescope, because Castel Gandolfo became too bright. With support of the Vatican Observatory Foundation, the Fred and Alice P. Lennon and Thomas J. Bannan Vatican Observatory Advanced Technology Telescope (VATT) was built on Mt. Graham (Arizona) in partnership with the University of Arizona.

The research performed there has demonstrated that it is a perfect instrument for CCD photometry. It obtained first light in 1993. It was successfully tested with borrowed CCD cameras, and with its new own "CCD26" 2k-camera began photometric runs, with a dedicated set of filters. The camera and filters were subsequently upgraded to give a full 12-arcmin field of view.

We can now say that the success with Schmidt photographic plates (Smriglio et al. 1986) and then of CCD KPNO observations was a milestone for the Vilnius System itself and for the Vatican Observatory (Fig. 13). F. Smriglio at the University "La Sapienza" in Rome also successfully used the Vilnius System with the CCD, and so did F. Vrba at the US Naval Observatory—Flagstaff Station (Boyle et al. 1996; Straižys et al. 1996b). Thus the Photometry Program of the Vatican Observatory with VATT and other telescopes grew in ever increasing value for stellar research.

The CCD imaging is far better than other methods for high quantum efficiency and it is linear response. The image surface variations can be removed by flat-fields what was shown by an important work of Laugalys et al. (2003). In Fig. 14, we show the CCD quality of a stellar profile: the radial profiles of three stars are Gaussian and the number of counts is much higher than with plates. This is the main reason of a great success of the CCD cameras used for astronomy.

We currently provide two types of CCD photometry: aperture photometry (Fig. 15) and PSF photometry (Fig. 16). Aperture photometry allows measuring the flux within the image of a star and the sky level on an annulus. The panoramic CCD camera effectively replaces hundreds of discrete photometers. This however may be not sufficient for instrumental convolution and crowded star fields, where the point-spread function (PSF model photometry) must be applied. The situation is shown below.

Obtaining photometry out of seven Vilnius filters provides us with "internal, instrumental system" data.

When we want to achieve 1% accuracy for good classification, we observe a calibrating well-known star field (M 67) as a pattern for flat-fielding. This extra step makes us more secure that our calculations fit well the new star fields. The adoption of more sophisticated methods of "mixing" data of separate runs is underway.

We pursue stellar analysis with observed photometric magnitudes. We can eliminate the reddening (Fig. 17) of our stars and classify them by means of the $Q$ parameters (Fig. 18).

To perform (preliminary = "SpCl") classification of stars we use the intrinsic Vilnius System stars, and also a database ("Catalog") of well-known "Vilnius" stars. We have created a software pipeline to do this sequence of tasks. Its results are shown in Fig. 18.

Thanks to the Vilnius System we can determine numerous astrophysical quantities (Straižys 2003):

- spectral class of a star
- its temperature
- absolute magnitude
- surface gravity
- metallicity
- color excess
- interstellar extinction
- distance

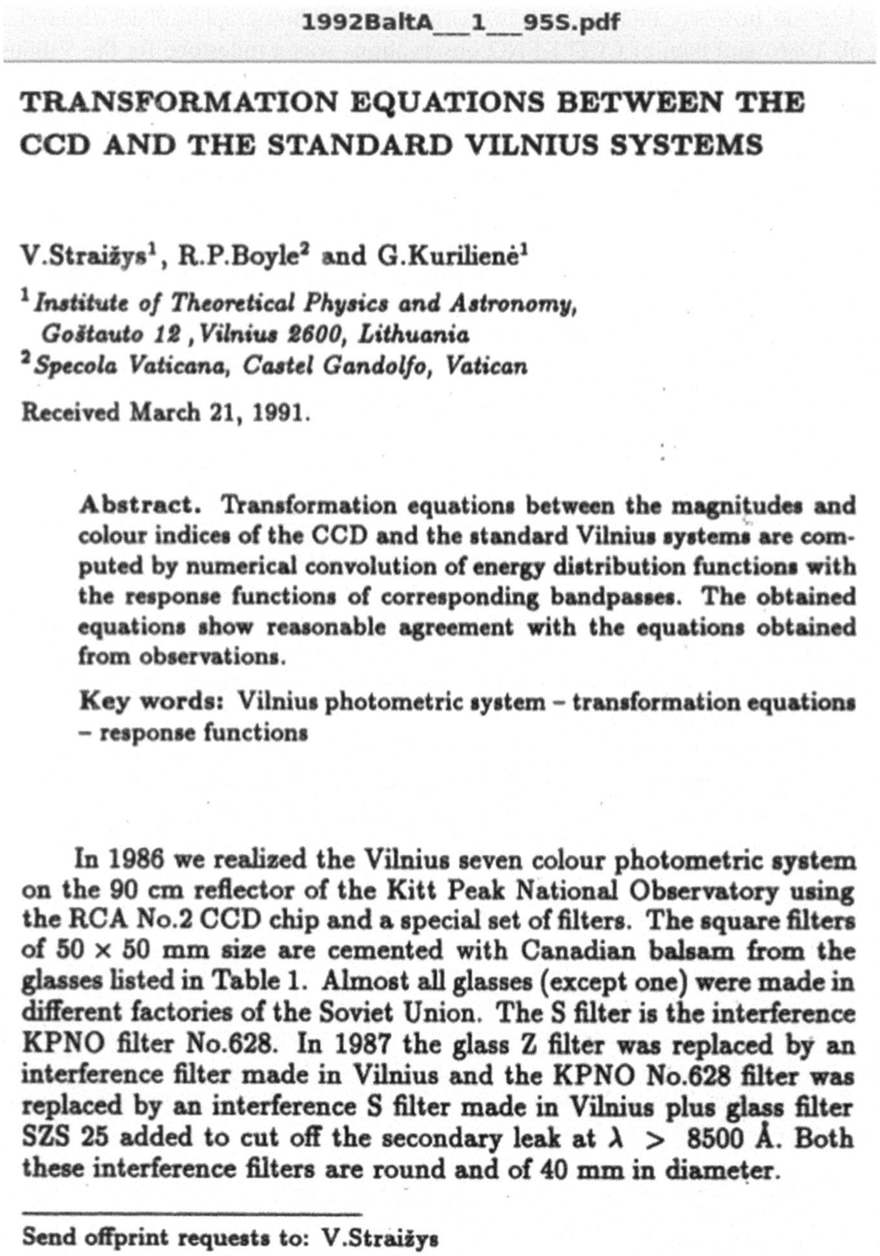

1992BaltA___1___95S.pdf

**TRANSFORMATION EQUATIONS BETWEEN THE CCD AND THE STANDARD VILNIUS SYSTEMS**

V.Straižys[1], R.P.Boyle[2] and G.Kurilienė[1]

[1] *Institute of Theoretical Physics and Astronomy, Goštauto 12, Vilnius 2600, Lithuania*
[2] *Specola Vaticana, Castel Gandolfo, Vatican*

Received March 21, 1991.

**Abstract.** Transformation equations between the magnitudes and colour indices of the CCD and the standard Vilnius systems are computed by numerical convolution of energy distribution functions with the response functions of corresponding bandpasses. The obtained equations show reasonable agreement with the equations obtained from observations.

**Key words:** Vilnius photometric system – transformation equations – response functions

In 1986 we realized the Vilnius seven colour photometric system on the 90 cm reflector of the Kitt Peak National Observatory using the RCA No.2 CCD chip and a special set of filters. The square filters of 50 × 50 mm size are cemented with Canadian balsam from the glasses listed in Table 1. Almost all glasses (except one) were made in different factories of the Soviet Union. The S filter is the interference KPNO filter No.628. In 1987 the glass Z filter was replaced by an interference filter made in Vilnius and the KPNO No.628 filter was replaced by an interference S filter made in Vilnius plus glass filter SZS 25 added to cut off the secondary leak at $\lambda > 8500$ Å. Both these interference filters are round and of 40 mm in diameter.

Send offprint requests to: V.Straižys

**Fig. 13** Vilnius system and the new CCD instrumentation (KPNO 1986), see Straižys et al. (1992)

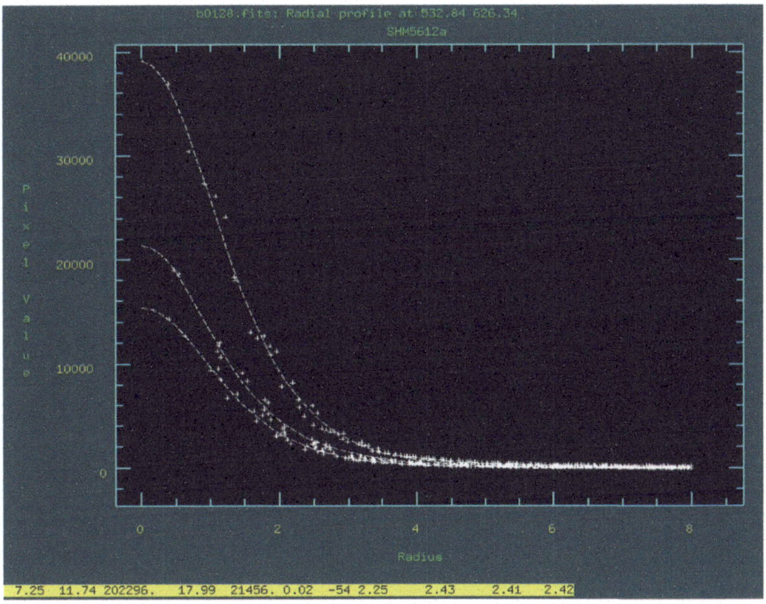


**Fig. 14** The linear quality of CCD stellar profile surpasses that obtained with photographic plates

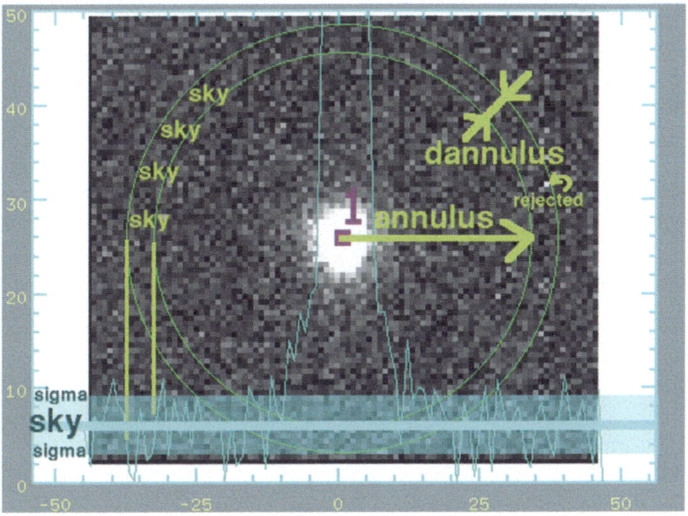


**Fig. 15** Aperture photometry concept. It is ideal for a single star or for star fields that are not crowded

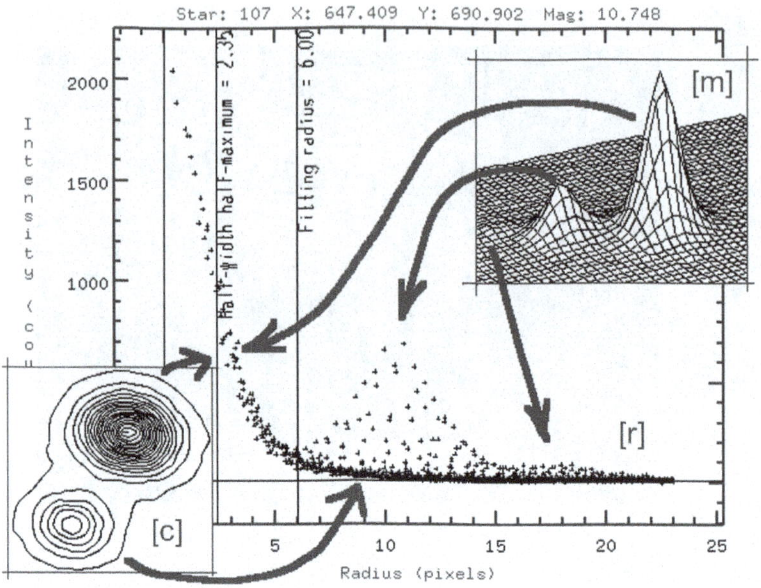


**Fig. 16** For crowded fields, stars must be deconvoluted by means of PSF

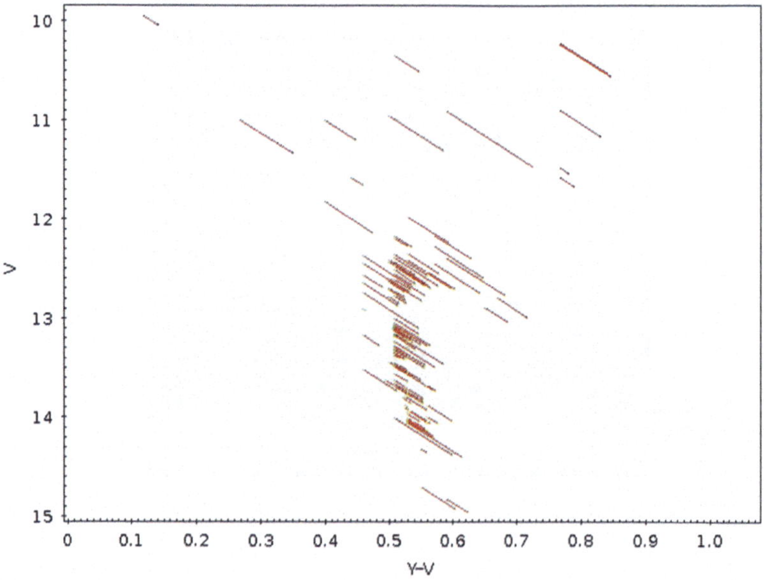


**Fig. 17** De-reddening of stars. The lines join the intrinsic and apparent positions on the color-magnitude diagram

| Star | | | Catalog | | | | comp. | #Q | mag |
|---|---|---|---|---|---|---|---|---|---|
| id | SpCl | dQ | id | SpCl | dQ | InfoSpCl: MK; Phot | dQx | used | V |
| w0040 | F8Z | 0.020 | BD+75___905 | F8Z | 0.017 | G0; F8_V | 0.008 | 10 | 13.809 |
| w0040 | F8Z | 0.020 | HD_224974 | F8Z | 0.021 | G0V; F8_IV | 0.009 | 10 | 13.809 |
| w0040 | F8Z | 0.020 | GSC_03180-00849 | F8V | 0.026 | _; G0_IV | 0.011 | 10 | 13.809 |
| w0040 | F8Z | 0.020 | NGC_2682_030 | F8IV | 0.022 | _; F8_IV-V | 0.013 | 10 | 13.809 |
| w0040 | F8Z | 0.020 | HD_283806 | F8Z | 0.027 | F8V; F8_V | 0.013 | 10 | 13.809 |
| w0040 | F8Z | 0.020 | BD-00__3409 | F8V | 0.024 | _; F-G | 0.014 | 10 | 13.809 |
| w0040 | F8Z | 0.020 | GSC_02543-00521 | F8Z | 0.02 | _; _ | 0.015 | 10 | 13.809 |
| w0040 | F8Z | 0.020 | GSC_01127-02429 | F8Z | 0.02 | _; F8_V | 0.015 | 10 | 13.809 |
| w0040 | F8Z | 0.020 | GSC_05693-05012 | F8Z | 0.021 | _; F8_V | 0.015 | 10 | 13.809 |
| w0040 | F8Z | 0.020 | CD-26_11316 | F8Z | 0.044 | _; F9;_UV_BR | 0.016 | 10 | 13.809 |
| w0042 | F8Z | 0.023 | HD__27991 | F8Z | 0.017 | F7V; _ | 0.010 | 12 | 13.495 |
| w0042 | F8Z | 0.023 | GSC_03180-02003 | F8V | 0.028 | _; F8_V | 0.011 | 12 | 13.495 |
| w0042 | F8Z | 0.023 | BD+37__2446 | F8V | 0.019 | F5; F7_V | 0.012 | 12 | 13.495 |
| w0042 | F8Z | 0.023 | BD+76___591 | F8V | 0.019 | F8; F8_V | 0.013 | 12 | 13.495 |
| w0042 | F8Z | 0.023 | HD__17206 | F8Z | 0.017 | F5/F6V; _ | 0.013 | 12 | 13.495 |
| w0042 | F8Z | 0.023 | NGC_2682_127 | F8V | 0.022 | _; F7/8_V | 0.014 | 12 | 13.495 |
| w0042 | F8Z | 0.023 | GSC_02543-00757 | F8V | 0.02 | _; _ | 0.015 | 12 | 13.495 |
| w0042 | F8Z | 0.023 | GSC_00023-00935 | F8V | 0.028 | _; F8_V | 0.015 | 12 | 13.495 |
| w0042 | F8Z | 0.023 | BD+75___905 | F8Z | 0.017 | G0; F8_V | 0.015 | 12 | 13.495 |
| w0042 | F8Z | 0.023 | GSC_05270-00050 | F8Z | 0.026 | _; _ | 0.016 | 12 | 13.495 |
| [illegible] | [illegible] | [illegible] | [illegible] | [illegible] | [illegible] | [illegible] | [illegible] | [illegible] | [illegible] |

**Fig. 18** One of our classification systems in terms of *Q-Q* parameters. The screen-shot of computer monitor shows the database-identifiers related to spectral and luminosity classes (SpCl) and related Catalog of Vilnius stars; the dQ and dQx denotes the quality of the classification

- many types of peculiarity
- ca. 10,000 photoelectric stars are studied
- many CCD star-fields are examined
- links to other (space) surveys are easy.

The VATT contribution to the Vilnius System survey includes:

- over 60 observation runs
- over 550 telescope pointings
- ca. 300 fields of view
- an international team
- continuous improvements on new data
- virtual observatory techniques (Fig. 19)

The sky view (Fig. 20) maps the star field observed for Vilnius photometry by VATT and a few other participating telescopes.

**Fig. 19** R.P. Boyle at the VATT

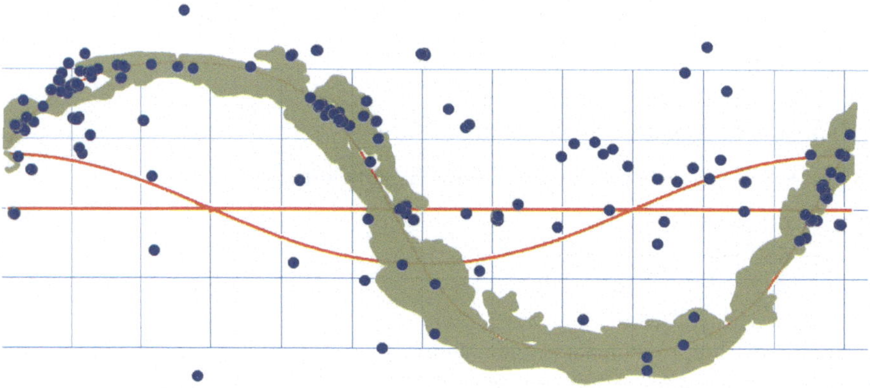

**Fig. 20** Our observation pointings on the sky map

**Acknowledgements** We thank our Referees for their remarks and additional informations.

## References


Bull. Vilnius Obs. (1970). The second conference on Vilnius intermediate band photometric system for three-dimensional classification of stars. *Vilnius Astronomijos Observatorijos Biuletenis, 28,* 3.

Boyle, R. P., Smriglio, F., Nandy, K., & Straižys, V. (1990). CCD photometry in the Vilnius photometric system. I. Region in Lyra. *Astronomy and Astrophysics Supplement Series, 84,* 1–10.

Boyle, R. P., Straižys, V., Vrba, F., Smriglio, F., Dasgupta, A. (1995). CCD photometry of the M67 in the Vilnius photometric system. In A. G. Davis Philip, K. Janes, A. R. Upgren (Eds.), *New developments in array technology and applications* (pp. 363–364).

Boyle, R. P., Vrba, F. J., Smriglio, F., Dasgupta, A. K., & Straižys, V. (1996). CCD observations in the Vilnius photometric system. *Baltic Astronomy, 5,* 231–237.

Corbally, C. J., & Boyle, R. P. (1987). A comparison between two-dimensional classifications from Vilnius photometry and those on the MK system. *Astronomy & Astrophysics, 186,* 114–119.

Gray, R. O., Corbally, C. J. (2009). *stellar spectral classification*. Princeton University Press.

Laugalys, V., Boyle, R. P., Kazlauskas, A., Vrba, F. J., Philip, A. G. Davis, & Straižys, V. (2003). Large-scale errors in CCD photometry of M67. *Baltic Astronomy, 12,* 497–502.

McCarthy, M. S. J. (1950). Fr. Secchi and Stellar Spectra. *Popular Astronomy, LVIII*(4), 574.

McCarthy, M. F. (1979). *Transactions of the International Astronomical Union, Series A, 17,* 83.

McCarthy, M. F., Philip, A. G. D., & Blaauw, A. (1977). Galactic structure in the direction of the polar caps. Joint discussion at the XVIth General Assembly of the I.A.U., Grenoble, 1976. *Highlights of Astronomy, 4,* 3–98.

Philip, A. G., Boyle, R. P., Straižys, V., & Kazlauskas, A. (2003). Setting up of the Stromvil system standards and standard regions. *Baltic Astronomy, 12,* 471–475.

Smriglio, F., Boyle, R. P., Straižys, V., Janulis, R., Nandy, K., MacGllivray, H. T., et al. (1986). Automated two-dimensional classification from multicolour photometry in the Vilnius system. *Astronomy & Astrophysics, Supplement Series, 66,* 181–190.

Smriglio, F., Dasgupta, A. K., & Boyle, R. P. (1994). Interstellar dust clouds in the direction of the globular cluster M71. *Astronomy and Astrophysics Supplement Series, 108,* 581–584.

Smriglio, F., Dasgupta, A. K., & Boyle, R. P. (1995). Interstellar dust clouds in the direction of the globular cluster M56. *Astronomy & Astrophysics, Supplement Series, 113,* 445–449.

Smriglio, F., Dasgupta, A. K., Nandy, K., & Boyle, R. P. (1990). A comparison of automated spectral classification using Vilnius photometry with MK classification. *Astronomy & Astrophysics, 228,* 399–402.

Straižys, V. (1963). *Vilnius Astronomijos Observatorijos Biuletenis, 6,* 1.

Straižys, V. (1970). The intermediate band photometric system for three-dimensional classification of stars. *Vilnius Astronomijos Observatorijos Biuletenis, 28*(6), 6.

Straižys, V. (1973). The Vilnius photometric system for three-dimensional classification of stars and determination of interstellar reddening. *Vilnius Astronomijos Observatorijos Biuletenis, 36,* 3.

Straižys, V. (1992). *Multicolor stellar photometry*. Tucson: Pachart Pub. House.

Straižys, V. (2003). Stellar photometry at the crossroads of the two centuries. *Baltic Astronomy, 12,* 433–446.

Straižys, V., Boyle, R., & Kuriliene, G. (1992). Transformation Equations between the CCD and the standard Vilnius system. *Baltic Astronomy, 1,* 95–106.

Straižys, V., Crawford, D. L., & Philip, A. G. D. (1996a). The Stromvil system: An effective combination of two medium-band photometric systems. *Baltic Astronomy, 5,* 83.

Straižys, V., Kazlauskas, A., Boyle, R. P., Vrba, F. J., & Smriglio, F. (1996b). Transformation equations between the standard and CCD Vilnius systems. II. The Flagstaff CCD system. *Baltic Astronomy, 5,* 165–180.

# Part III
# Galaxies

# Clusters of Galaxies… and Some Jellyfishes in the Sky

**Alessandro Omizzolo**

**Abstract** In this paper i briefly summarize the works of the Padova research group WINGS to which i belong. Two main topics are discussed: the WINGS project and a subproject about a particular kind of galaxies: the Jellyfish galaxies. I present the results from the U-band observations done at the Bok Telescope at Kitt Peak in Arizona, the OmegaWINGS survey and the first data about jellyfish galaxies.

**Keywords** Galaxies · Cluster of galaxies · Photometry

## From Clusters to Cosmology

Galaxy clusters, the most massive collapsed structures in the Universe, are the tail of a continuum distribution of halo masses, and the most extreme environments where galaxy formation has proceeded at an accelerated rate compared to the rest of the Universe. Clusters have been a test bed for studies of galaxy formation and evolution, uncovering trends that several years later have also been found in the field. They are a repository for galaxies that have been shaped in lower halo-mass environments, but are also the sites where essentially all environmental effects are thought to take place, from strangulation to ram-pressure stripping, and even galaxy merging. As peaks in the matter distribution, galaxy clusters host those galaxies that have formed first and in the most extreme primordial conditions. At the same time they are the sites where hierarchical growth is most evident, as in the case of the brightest cluster galaxies. There is no better place than rich clusters in the low-$z$ Universe to find and study the descendants of the most massive galaxies observed at high-$z$.

The powerful images released by HST and by the largest ground based telescopes increased the interest in the field of observational cosmology and a great

A. Omizzolo (✉)
Specola Vaticana (Vatican Observatory) and OAPD-INAF, vicolo dell'osservatorio n. 5, 35122 Padova, Italy
e-mail: alessandro.ommizolo@oapd.inaf.it

 
G. Gionti, S.J. and J.-B. Kikwaya Eluo, S.J. (eds.), *The Vatican Observatory, Castel Gandolfo: 80th Anniversary Celebration*, Astrophysics and Space Science Proceedings 51, https://doi.org/10.1007/978-3-319-67205-2_7

amount of papers have been published in these last 15 years studying the distant clusters of galaxies. Galaxy evolution, dark matter, gravitational lensing are only some of the topics of these papers (Biviano 2000; Suto 2000; Borgani and Guzzo 2001; Bohringer 2000; Voit 2005; Diaferio et al. 2008).

However, little interest was reserved to clusters in the local Universe and so there was a gap in the knowledge of these large structures in our universe.

Clusters of galaxies are dense peaks in the galaxy distribution (see Figs. 1, 2 and 3) and therefore appropriate sites to look for changes in the properties of the galaxies. They can be therefore used to trace the evolution of the systems themselves as well as that of the galaxies in them. But our present knowledge of the systematic properties of galaxies in nearby clusters remains surprisingly limited, with Virgo, Coma and Fornax as the main references.

**Fig. 1** The cluster Abell 1983. The image is the final result of the composition of three images in the B, V, U band. The distance is z = 0.0436 corresponding to about 0.607 Gly

**Fig. 2** The cluster Z8338; z = 0.0473 corresponding to about 0.65 Gly

In the range $0.4 \leq z \leq 0.5$ it has been found that spirals are a factor of 2–3 more abundant and S0 galaxies are proportionally less abundant than in nearby clusters, while the fraction of ellipticals is already as large or larger. This implies significant morphological transformations occurring rather recently.

Also the cluster type plays a role in determining the relative occurrence of S0 and elliptical galaxies at a given redshift.

To fill this gap our group started about 10 years ago a project to study a sample of X-ray selected nearby clusters of galaxies and in particular to study the cosmic variance of the cluster properties and their populations in a systematic way.

**Fig. 3** The cluster Abell 1831; z = 0.0615 corresponding to about 0.85 Gly

## The WINGS Project

Our project, whose acronym is WINGS standing for Wide-field Nearby Galaxy-cluster Survey, is a wide field multi-wavelength imaging and spectroscopic survey of galaxies in 77 nearby clusters (Fasano et al. 2006).

The sample was extracted from the ROSAT catalog of X-ray emitting clusters, with only two constraints: the redshift $0.04 < z < 0.07$ and the distance from the galactic plane $|b| \geq 20°$.

The global goal of the WINGS project is the systematic study of the local cosmic variance of the cluster population and of the properties of cluster galaxies as a function of cluster properties and local environment. This data collection allows the definition of a local, "zero-point" reference against which to gauge the cosmic evolution when compared to more distant clusters.

The core of the project consists of wide-field optical imaging of the selected clusters in the *B* and *V* bands. We have also completed a multi-fiber, medium-resolution spectroscopic survey for the majority of the clusters in the master sample. The imaging and spectroscopy data were collected using, respectively, the WFC@INT and WYFFOS@WHT in the northern hemisphere, and the WFI@MPG and 2dF@AAT in the southern hemisphere. In addition, a NIR (*J*, *K*) survey of ~50 clusters and an $H_\alpha$ + *U* survey of some 17 clusters have been done with the WFCAM@UKIRT and WFC@INT and 90prime@BOK, respectively, while a very-wide-field optical survey is ongoing with OmegaCam@VST.

In this way we want to elaborate a statistically meaningful, high quality database of the properties of nearby clusters of galaxies and of the galaxies that populate them. In broad terms, the goals of the project are to characterize the global properties of clusters taken as systems, and those of their member galaxies. Among the former, besides the already existing data on the X-ray luminosity, we include their total luminosity and size, the velocity dispersion, the presence of substructures and the cluster scaling relations (Marmo et al. 2004; Pignatelli et al. 2003, 2004).

Regarding the member galaxies, our primary goals are to analyse the variance of the morphological fractions (E/S0/S/Irr), their distribution in the clusters and the morphology-density relation. The analysis of the colours and the spectral information provides the data necessary to retrace the star formation history of galaxies in nearby clusters.

The WINGS project was designed to cover all these topics. Originally it was planned as a wide-field optical (*B*, *V*) imaging survey. This is the core of the project, hereafter called WINGS-OPT.

In addition, other surveys were designed and carried out to complement the characterization of the cluster galaxies. The already completed WINGS-SPE survey consists of multi-fibre spectroscopy of galaxies in 51 clusters from the master WINGS sample, obtained with the WYFFOS@WHT and the 2dF@AAT spectrographs over the same area covered by the optical imaging (34′ × 34′). The spectra cover the range 3800–7000 Å (WYFFOS) and 3600–8000 Å (2dF), with dispersions of 3 and 9 Å, respectively, for the galaxies with $V < 20$ (between 100 and 300 per cluster). This limit is 1.5 and 2.0 mag deeper than the 2dF and Sloan surveys, respectively (Fritz et al. 2007, 2011, 2014; Cava et al. 2009).

Three more follow-up surveys of clusters in the WINGS sample have been done. The first one is a NIR (WINGS- NIR: *J* and *K*-bands) imaging survey, with the Wide-Field Camera at the 3.8 m UKIRT telescope (Valentinuzzi et al. 2009, 2011). Data for ~50 clusters were obtained, useful at providing an estimate of the stellar mass of galaxies, as well as at constraining the spectral energy distribution of galaxies in the clusters. The other ones are $H_\alpha$ and U-broad-band surveys (WINGS-HAL and WINGS-UV, respectively), with the WFC@INT camera and purpose-defined narrow-band filters (for the WINGS-HAL survey), imaging ~1 square degree of 10 WINGS clusters. Finally, a very-wide-field ( ~ 1 square degree) optical survey (WINGS-VWF), with the ESO-VST telescope, equipped with OmegaCam, is completed in the B and V filters and ongoing in the u-band.

For all galaxies down to the limit of detectability we have extracted the position, size, concentration, average flattening and orientation, as well as the integrated and aperture photometry in the two observed bands, *B*, *V* (Moretti et al. 2014). For a subsample of large galaxies we have also obtained detailed surface photometry (luminosity and geometrical profiles) and global structural parameters (total magnitudes, effective radii, ellipticity and Sérsic index) using our automatic surface photometry tool GASPHOT (Pignatelli et al. 2006; D'Onofrio et al. 2014). Finally, morphological type estimates of the same subsample of galaxies, compared and calibrated with visual classifications, were automatically obtained with the purposely-written tool MORPHOT (Fasano et al. 2005).

The WINGS clusters span a wide range in X-ray luminosities (log $L_X$[0.1 −2.4 keV] = 43.2−44.7), ranging from $\sim 8 \times 10^{13}$ to $>10^{15}$ $M_\odot$ (Reiprich and Bohringer 2002), as well as in optical properties such as Abell richness and Bautz-Morgan type.

The WINGS-OPT survey should to be able to sample the luminosity function of clusters down to the dwarf galaxies ($M_V \sim -14$). The depth is sufficient to allow a reliable surface photometry (*S*/*N* ratio ≈ 4.5 per square arcseconds) down to a surface brightness of $\mu_V \sim 25$ mag arcsec$^{-2}$ (Varela et al. 2009).

To get the most and the best from this huge amount of data a dedicated software was developed by some of us, namely GASPHOT to do automatic photometry, MORPHOT (Fasano et al. 2012) to do automatic morphologic classification, and SPSs Code to analize the spectra and to trace star formation histories.

## The U-Band Survey Done at 90prime@BOK

The accessibility to the facility of the Steward Observatory allowed us to apply for time at the Bok telescope to take advantage of the wide field imager 90′ at the Bok telescope. We did so because many observations have proved that the cluster outskirts are essential to understand galaxy transformations. In particular several recent works have shown that in the local Universe the correlation between star formation activity and local density extends to very large clustercentric radii, well beyond the cluster central regions.

So accessing the Bok imager we combined two necessities of our survey: both the study of the outskirts of the clusters and the star formation (Omizzolo et al. 2014). In fact, the integrated spectrum of a galaxy is more and more dominated by young stars going to shorter wavelengths and the U-band data are far more sensitive to any current or recent star formation than the other broad-bands available. The following pictures (Figs. 1, 2 and 3) show three clusters imaged at the Bok Telescope.

## The OmegaWINGS Survey

OmegaWINGS is a photometric and spectroscopic survey of 46 nearby galaxy clusters (Gullieuszik et al. 2015a, b). The photometric part is based on OmegaCAM imaging with the VLT Survey Telescope; the field of view of one square degree covers the whole virial region and beyond, extending up to the infall regions of each OmegaWINGS cluster. The spectroscopic survey has provided about 30,000 spectra with AAOmega on the Australian Astronomical Telescope. OmegaWINGS thus provides an extremely detailed and complete view of galaxy populations in clusters.

OmegaWINGS targeted clusters were randomly selected from the 57 WINGS clusters that can be observed from the VST (<+20°). We obtained service mode B and V imaging for 46 clusters.

In Fig. 4 we show an example of the wide field image from OmegaCam of the cluster A2399. The grey square is the field of the old WINGS image, the circle shows the radius of the cluster (R200), the colour images zooming into the three selected areas were obtained combining OmegaWINGS U, B, V band images.

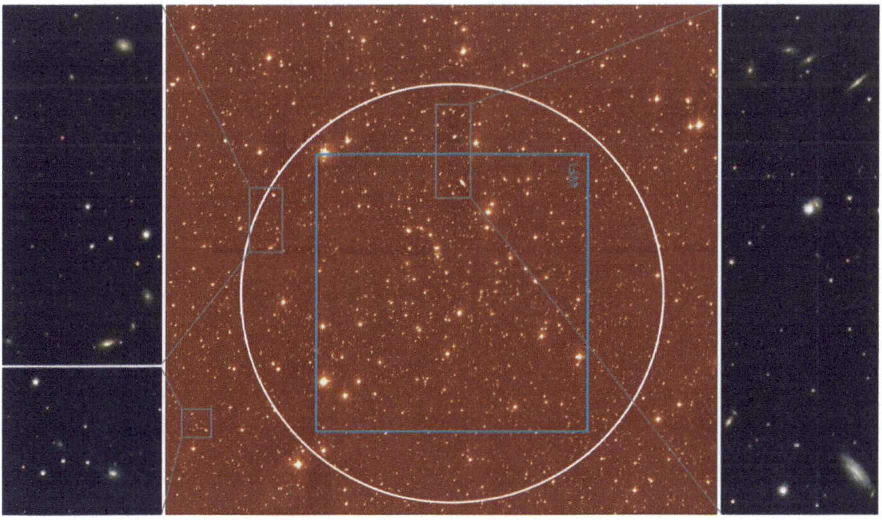

**Fig. 4** A wide field image of the cluster A2399 from Omegacam

## Jellyfish Galaxies Candidates

In order to unveil the physical drivers of galaxy evolution, it is crucial to study the processes of gas acquisition and loss. Gas is the fuel for star formation (SF) and a sensitive tracer of environmental effects.

Gas loss from galaxies can be caused by mechanisms internal to galaxies themselves, such as galactic winds due to star formation or an AGN (e.g. Veilleux et al. 2005; Ho et al. 2014; Fogarty et al. 2012). In addition, several external mechanisms that can potentially impact on a galaxy gas content have been proposed (Boselli and Gavazzi 2006; De Lucia 2010). Among those not directly affecting the galaxy stellar component, there are ram pressure stripping from the disk due to the interaction between the galaxy ISM and the intergalactic medium (IGM, Gunn and Gott 1972), and the removal of the hot gas halo surrounding the galaxy (the so-called "strangulation") either via ram pressure or via tidal stripping by the halo potential (Larson et al. 1980; Balogh et al. 2000).

Among those processes that affect both gas and stars, instead, there are strong tidal interactions and mergers (expected to be more common in groups, Barnes and Hernquist 1992), tidal effects of the cluster as a whole (Byrd and Valtonen 1990) and "harassment", i.e. the cumulative effect of several weak and fast tidal encounters, expected to be more efficient in galaxy clusters (Moore et al. 1996).

Galaxies that are being stripped of their gas can sometimes be recognized from their optical appearance. Extreme examples of stripped galaxies are the so-called "jellyfish galaxies", that exhibit tentacles of debris material with a characteristic jellyfish morphology. We have conducted the first systematic search for galaxies that are being stripped of their gas at low-z ($z = 0.04 - 0.07$) in different environments, selecting galaxies with varying degrees of morphological evidence for stripping. We have visually inspected the B and V-band images and identified 344 candidates in $\sim$70 galaxy clusters of the OMEGAWINGS + WINGS sample and 75 candidates in groups and lower mass structures in the PM2GC sample. The PM2GC is a galaxy catalogue representative of the general field population in the local universe. The galaxy environments is characterized by identifying galaxy groups at $0.04 \leq z \leq 0.1$ with a Friends-of-Friends (FoF) algorithm using a complete sample of 3210 galaxies brighter than MB = −18.7 taken from the Millennium Galaxy Catalogue (MGC, Liske et al. (2003)), a 38 degree$^2$ photometric and spectroscopic equatorial survey.

We have complied an atlas of stripping candidates and carried out a first analysis of their environment and their basic properties, such as morphologies, star formation rates and galaxy stellar masses. Candidates are found in all clusters and at all clustercentric radii, and their number does not correlate with the cluster velocity dispersion $\sigma$ or X-ray luminosity $L_X$. Interestingly, convincing cases of stripping

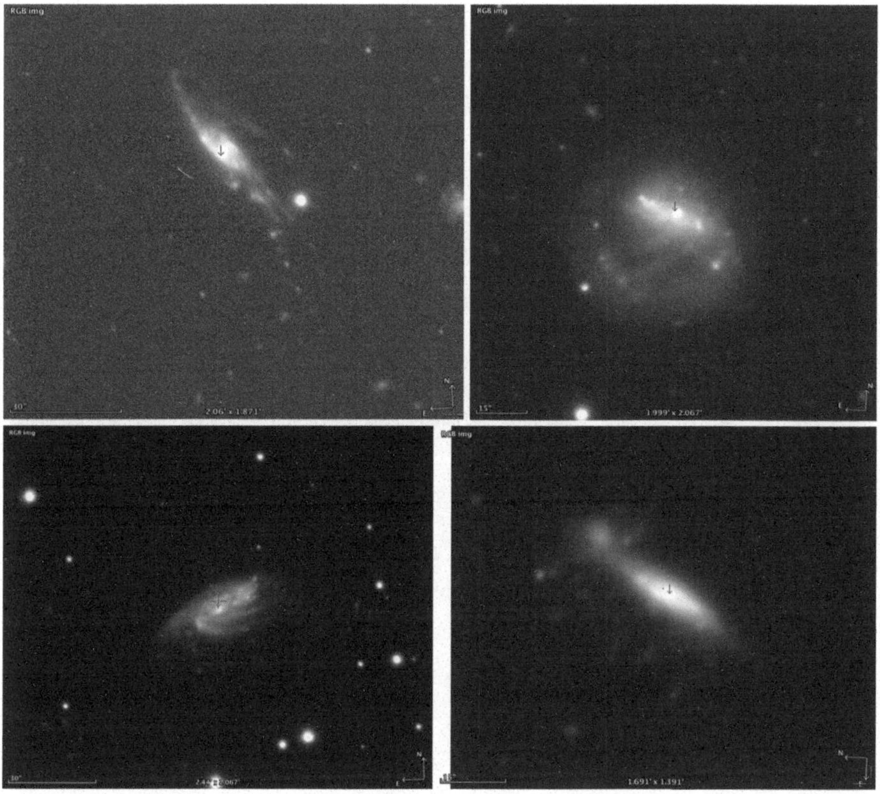

**Fig. 5** Some examples of jellyfish galaxies found in our OmegaWINGS survey

candidates are also found in groups and lower mass haloes ($10^{11}$–$10^{14}$ $M_{\odot}$). All the candidates are disky, have stellar masses ranging from $\log M/M_{\odot}$ <9 to >11.5 and the majority of them form stars at a rate that is on average a factor of 2 higher compared to non-stripped galaxies of similar mass. The few post-starburst and passive candidates have weak stripping evidence. We conclude that the stripping phenomenon is ubiquitous in clusters and could be present even in groups and low mass haloes. Further studies will reveal the physics of the gas stripping and clarify the mechanisms at work.

In Fig. 5 we show some examples of the most spectacular jellyfish galaxies found in our OmegaWINGS survey.

Fig. 5 (continued)

## References


Balogh, M. L., et al. (2000). *ApJ, 540,* 113–121.

Barnes, J. E., & Hernquist, L. (1992). *ARAA, 30,* 705.

Biviano, A. (2000). Constructing the universe with clusters of galaxies. In F. Durret & D. Gerbal (Eds.), *IAP 2000 Meeting*, Paris, France, July 2000.

Bohringer, H. (2000). X-ray Astronomy 2000. In R. Giacconi, S. Serio, & L. Stella (Eds.), *ASP Conference Proceeding* (Vol. 234). San Francisco: Astronomical Society of the Pacific.

Borgani, S., & Guzzo, L. (2001). *Nature, 409*(6816), 39–45.

Boselli, A., & Gavazzi, G. (2006). *PASP, 118,* 517.

Byrd, G., & Valtonen, M. (1990). *ApJ, 350,* 89.

Cava, A., Bettoni, D., Poggianti, B. M., Couch, W. J., Moles, M., Varela, J., et al. (2009). WINGS-SPE spectroscopy in the WIde-field nearby galaxy-cluster survey. *Astronomy & Astrophysics, 495*(3), 707–719.

De Lucia, G. (2010). arXiv 1012.3326.

Diaferio, A., et al. (2008). *Space Science Reviews, 134*(1–4), 7–24.

D'Onofrio, M., Bindoni, D., Fasano, G., Bettoni, D., Cava, A., Fritz, J., et al. (2014). Surface photometry of WINGS galaxies with GASPHOT. *A&A, 572,* A87. (eprint arXiv:1410.2441).

Fasano, G., Marmo, C., Varela, J., D'Onofrio, M., Poggianti, B. M., Moles, M., et al. (2005). WINGS: a WIde-field nearby Galaxy-cluster survey. *Astronomy & Astrophysics, 445*(3), 805–817.
Fasano, G., Marmo, C., Varela, J., D'Onofrio, M., Poggianti, B. M., Moles, M., et al. (2006). WINGS: A WIde-field Nearby Galaxy-cluster Survey. I. Optical imaging. *A&A, 445,* 805.
Fasano, G., Vanzella, E., Dressler, A., Poggianti, B. M., Moles, M., Bettoni, D., et al. (2012). Morphology of galaxies in the WINGS clusters. *MNRAS, 420,* 926. (eprint arXiv:1109.2026).
Fogarty, L. M. R., et al. (2012). *ApJ, 761,* 169.
Fritz, J., Poggianti, B. M., Bettoni, D., Cava, A., Couch, W. J., D'Onofrio, M., et al. (2007). A spectrophotometric model applied to cluster galaxies: the WINGS dataset. *A&A, 470,* 137.
Fritz, J., Poggianti, B. M., Cava, A., Moretti, A., Varela, J., Bettoni, D., et al. (2014). WINGS-SPE. III. Equivalent width measurements, spectral properties, and evolution of local cluster galaxies. *A&A, 566,* A32. (eprint: arXiv:1402.4131).
Fritz, J., Poggianti, B. M., Cava, A., Valentinuzzi, T., Moretti, A., Bettoni, D., et al. (2011). WINGS-SPE II: A catalog of stellar ages and star formation histories, stellar masses and dust extinction values for local clusters galaxies. *A&A, 526,* A45.
Gullieuszik, M., Poggianti, B. M., Fasano, G., Zaggia, S., Paccagnella, A., Moretti, A., et al. (2015a). OmegaWINGS: OmegaCAM@VST observations of WINGS galaxy clusters. *A&A, 581,* A41. (eprint: arXiv:1503.02628).
Gullieuszik, M., Poggianti, B. M., Fasano, G., Zaggia, S., Paccagnella, A., Moretti, A., et al. (2015b). OmegaWINGS: A VST survey of Nearby galaxy clusters. *The Messenger, 160,* 17.
Gunn, J. E., & Gott, J. R. (1972). *ApJ, 176,* 1–19.
Ho, I.-T., et al. (2014). *MNRAS, 3894,* 3910.
Larson, R. B., et al. (1980). *ApJ, 237,* 692–707.
Liske, J., et al. (2003). *MNRAS, 344,* 307.
Marmo, C., et al. (2004). Clusters of galaxies: Probes of cosmological structure and galaxy evolution, from the Carnegie observatories centennial symposia. In J. S. Mulchaey, A. Dressler, & A. Oemler (Eds.), *Carnegie Observatories Astrophysics Series, 2004*. Pasadena: Carnegie Observatories.
Moore, B., et al. (1996). *Nature, 376,* 613.
Moretti, A., Poggianti, B. M., Fasano, G., Bettoni, D., D'Onofrio, M., Fritz, J., et al. (2014). WINGS data release: A database of galaxies in nearby clusters. *A&A, 564,* A138. (eprint arXiv:1403.1408).
Omizzolo, A., et al. (2014). *A&A, 561,* A11.
Pignatelli, E., Marmo, C., Poggianti, B. M., Bettoni, D., Fasano, G., Varela, J., et al. (2003). WINGS: The first results. *MmSAI, 3,* 265.
Pignatelli, E., Marmo, C., Poggianti, B. M., Bettoni, D., Fasano, G., Varela, J., et al. (2004). WINGS: A photometric, morphological and spectroscopic library of nearby cluster galaxies. *MmSAI, 75,* 214.
Pignatelli, E., et al. (2006). *A&A, 446,* 373–388.
Reiprich, T., Bohringer, H. (2002). *ESO Astrophysics Symposia*.
Suto, Y. (2000). Constructing the Universe with Clusters of Galaxies. In F. Durret & D. Gerbal (Eds.), *IAP 2000 Meeting*, Paris, France, July 2000.
Valentinuzzi, T., Woods, D., Fasano, G., Riello, M., D'Onofrio, M., Varela, J., et al. (2009). WINGS: a WIde-field nearby Galaxy-cluster survey. *Astronomy & Astrophysics, 501*(3), 851–864.
Valentinuzzi, T., Poggianti, B. M., Fasano, G., D'Onofrio, M., Moretti, A., Ramella, M., et al. (2011). The red-sequence of 72 WINGS local galaxy clusters. *A&A, 536,* A34. (eprint arXiv:1109.4011).
Varela, J., D'Onofrio, M., Marmo, C., Fasano, G., Bettoni, D., Cava, A., et al. (2009). WINGS: A WIde-field Nearby Galaxy-cluster Survey. II. Deep optical photometry of 77 nearby clusters. *A&A, 497,* 667.
Veilleux, S., et al. (2005). *ARA&A, 43,* 769–826.
Voit, G. M. (2005). *Reviews of Modern Physics, 77*(1), 207–258.

# Stellar Haloes of Galaxies

**Richard D'Souza**

**Abstract** The stellar halo of a galaxy offer us a unique way of probing its merger history. In this paper, I outline a method of studying the average properties of the stellar haloes of galaxies as a function of stellar mass and galaxy type by stacking a large number of similar galaxies together from the large all-sky Sloan Digital Sky Survey (SDSS). By stacking nearly $\sim$10,000 galaxies together in bins of stellar mass and galaxy type, we can study the surface brightness profiles of these galaxies to a depth of 31 magnitudes/arcsec$^2$ in the r-band and out to a distance of nearly 100 kpc away from the center of a galaxy. Measuring the amount of accreted stellar material in the haloes of galaxies also allows us to test our present theories of galaxy formation.

## Introduction

If one would take an image of a typical galaxy through an optical telescope, one would observe that the galaxy is very bright at its center and that its surface brightness decreases gradually towards its outer parts. Beyond a certain limiting radius, the CCD detector of the telescope would not register any light. However, if it were possible to integrate deeper, one would find that the light of the galaxy extends much further out, typically more than 4 or 5 times the normal limiting radius, depending upon how long one observes. This extended optical low surface brightness outer region arises from stars which surrounds a galaxy and is generally called it's stellar halo. Owing to their extremely low surface brightness (a 100 times fainter than the brightness of the night sky), the stellar haloes of galaxies are very difficult to observe.

If one had the ability to further resolve this outer low surface brightness region, as it is possible for nearby galaxies with the Hubble Space Telescope, one would see that the stellar halo of a galaxy is made up primarily of many streams and shells of stars. This indicates that the stellar halo of a galaxy is made up of the remnants of smaller

R. D'Souza (✉)
Specola Vaticana, Padova, Italy
e-mail: rdsouza@specola.va



G. Gionti, S.J. and J.-B. Kikwaya Eluo, S.J. (eds.), *The Vatican Observatory, Castel Gandolfo: 80th Anniversary Celebration*, Astrophysics and Space Science Proceedings 51, https://doi.org/10.1007/978-3-319-67205-2_8

galaxies that are tidally disrupted. This is consistent with our current view of galaxy formation: that galaxies grow through hierarchal merging, where smaller galaxies merge together to form larger galaxies. Stellar haloes are thus formed as a natural byproduct of galaxy formation. The stellar haloes of galaxies are made up of mainly stars which have been accreted through the tidal disruption of the smaller galaxies.

Despite containing only a small fraction of the total light of the galaxy, the stellar halo arguably contains the most useful information about the evolutionary history of a galaxy. This is because stars in the halo provide a fossil record of the assembly history of the galaxy through their chemical abundances and motions. The stars in the stellar haloes retain in their atmospheres a record of the chemical elements of the environment in which they were born. Additionally, the long relaxation time of stars in the halo allows them to retain some record of their origins in phase-space (Figs. 1 and 2).

In principle, it would be possible to decode the accretion history of a galaxy by studying its stellar halo. Over the last few years, we have made a lot of progress in understanding the various processes which are responsible for the formation of the stellar halo. Much of this has been driven by advancements in techniques of numerical simulations of galaxies. For instance, these simulations have indicated to us that the majority of the stellar halo was built through the merger one or more massive satellites about 9–10 Gyrs ago. On the other hand, the surviving satellites

**Fig. 1** The stellar halo with a Sagitarius-like stream of Messier 63 taken through a small aperture telescope from Martínez-Delgado et al. (2010), The Astronomical Journal, 140, 962

**Fig. 2** The stellar halo of an early-type (elliptical) galaxy NGC0474 taken through the Canada–France–Hawaii 3.5 m Telescope

of a galaxy have been accreted later, about 4–5 Gyrs ago. However, it still remains a difficult task to decode the accretion history of a galaxy from its stellar halo. While this may be achievable in the coming years, we could still use the stellar haloes of galaxies to test our theories of galaxy formation.

## Testing Galaxy Formation Models

A galaxy is a mixture of a large set of stellar populations. Not only are groups of stars born at different times in the galaxy's history, but a sizable amount of stars are often formed in neighboring smaller galaxies and accreted into the main galaxy. Therefore, it is natural to divide the stellar populations into stars that are born with in the galaxy itself (in situ stars) and stars that have been born elsewhere and have been accreted into the main galaxy during the course of its life time (accreted stars). The fraction of accreted stars of a galaxy give us an important indicator of its formation processes. Simulations indicate that disk galaxies like our Milky Way have low accreted fractions (between 0 and 5%). On the other hand, elliptical galaxies which often have suffered a larger number of mergers in the past have higher accreted fractions (between 50 and 70%).

However, these theories of galaxy formation must be confronted and verified by observations of the stellar haloes of galaxies. In particular, observational constraints of the stellar haloes of galaxies would help us ask the following set of questions: (a) what are the average properties of the stellar haloes of galaxies as a function of galaxy type? How far do they extend, and what is the average radial profile of the galaxy? (b) What is the percentage of accreted stars in a galaxy as a function of its type? (c) How much of the outer stellar material of galaxies do we miss out in large all-sky surveys? Answers to such questions require not only deep integrated light observations of the low surface brightness regions around galaxies, but also that these observations have to repeated for a large number of galaxies representative of the observable universe.

Observations of the stellar halo of a galaxy are expensive and require a long integration time ($\sim$ a few hours). Practically, deep observations are done by co-adding (stacking) a large number of short individual exposures. Co-adding a large number of images of a galaxy decreases the noise and increases the signal-to-noise ratio in the final stacked image. In low surface brightness observations, smaller telescopes often fare much better than larger ones. The intervening structural elements of larger telescopes contributes to the scattering of light, making it difficult to distinguish out the low surface brightness regions. However, smaller telescopes have an intrinsic disadvantage in the amount of light they can collect leading to much larger integration times. Small aperture telescopes often need integration times in excess of 12 h or more.

## Stacking Galaxies Together

A possible solution to derive the average properties of the stellar haloes of galaxies would be to co-add images of many different but similar-type galaxies. Similar to co-adding different images of a single galaxy together, this technique allows us to derive the average stellar halo of a particular galaxy type using currently available all-sky surveys. The Sloan Digital Sky Survey (SDSS) which has mapped a third of the night sky and measured the stellar masses and average properties of nearly 500,000 local galaxies, would be ideal for such purposes. By co-adding SDSS images of a large number of similar galaxies, we could detect and derive the *average* properties of their stellar haloes. For example, by stacking about $\sim$10,000 SDSS elliptical galaxies together, we could extend the average radial surface brightness profile of a galaxy to a level which is 100 times fainter (31 magnitudes/arcsec$^2$ in the SDSS r-band) and up to a radial extent which is 4–5 times larger than from a single galaxy. Stacking such a large number of galaxy images together allows us to detect the stellar halo of a galaxy out to a typical distance of a 100 kpc (Figs. 3 and 4).

Although such a stacking exercise have been attempted in the past, this was the first time that we stacked galaxies together as a function of their stellar mass and galaxy type, allowing us to derive the average radial properties of the stellar halo for a wide range of galaxies—from small disk galaxies up to large elliptical galaxies at

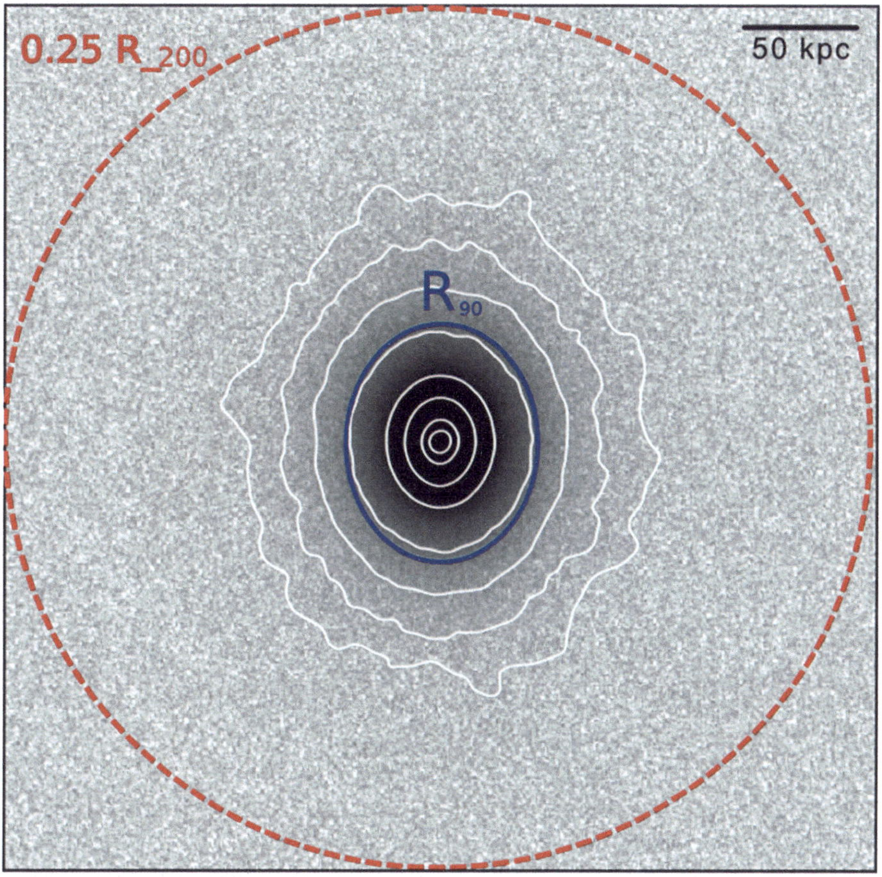


**Fig. 3** Ellipticity contours drawn on the stacked image of large early-type galaxies. The over-plotted blue and red lines indicate $\mathbf{R}_{90}$ (radius which encloses 90% of the light of the galaxy) and 0.25 $\mathbf{R}_{200}$ (radius of the dark matter halo) respectively

the center of galaxy groups and clusters. This exercise was possible at this stage because of the maturity and uniformity of the whole completed SDSS data set, and our present ability (nearly 10 years later) to understand and deal with the systematic problems with the SDSS data set. A common way to characterize the stellar halo of a galaxy is through its radial surface brightness profile. It measures how the average intensity of light decreases outwards with distance.

First of all, these deep radial surface brightness profiles of galaxies indicate to us how much light we actually miss from all-sky surveys. We find that for smaller galaxies we miss very little of the outer light of the galaxy. For larger elliptical galaxies, surveys like the SDSS can miss up to nearly 15–20% of the light which is present in the outer low surface brightness region of the galaxy. This additional stellar material at the outer rim of galaxies adds to the total missing stellar budget of the Universe. It increases the stellar mass density of larger galaxies (greater than

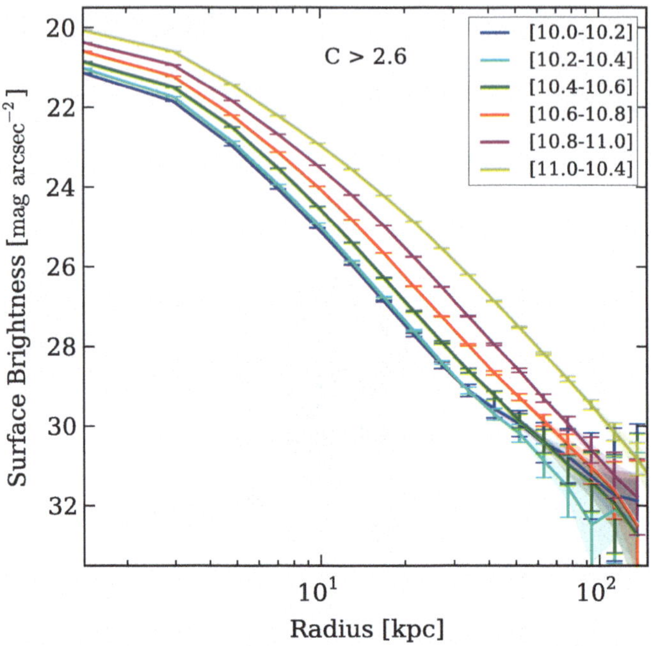


**Fig. 4** Radial surface brightness profiles of galaxy stacks for early-type galaxies (ellipticals) for various stellar masses D'Souza et al. (2014)

3 times the mass of the Milky Way) by nearly 3.36 times the traditional estimates of the same from the SDSS data set D'Souza et al. (2015).

Secondly, the shape of the radial surface brightness profile encodes important information about formation of the galaxy and its stellar halo. It also encodes information about the amount of stellar material that has been accreted by the galaxy. For example, galaxies with larger stellar haloes will have a more extended radial surface brightness profile which falls gently with radius. Galaxies with a smaller stellar halo will have a surface brightness profile which falls sharply with increasing distance (Fig. 5).

To decode this information, we need to measure reliably the accreted stellar populations in the outer parts of the galaxy and extrapolate it towards the inner parts. It has been long argued from theoretical considerations that the orbits of insitu-stars and accreted stars would be different from each other, leading to a separation of components in the radial surface brightness profile. The outer parts, at large galacto-centric distances would be made up predominately of accreted stars.

Furthermore, the amount of accreted stellar material in the outer parts would be proportional to the amount of total accreted stellar material. With the help of models from numerical simulations, it would be possible to extrapolate from the amount of accreted stellar material in the outer parts to the total amount of accreted stellar material. According to dynamical friction arguments and borne out by numerical

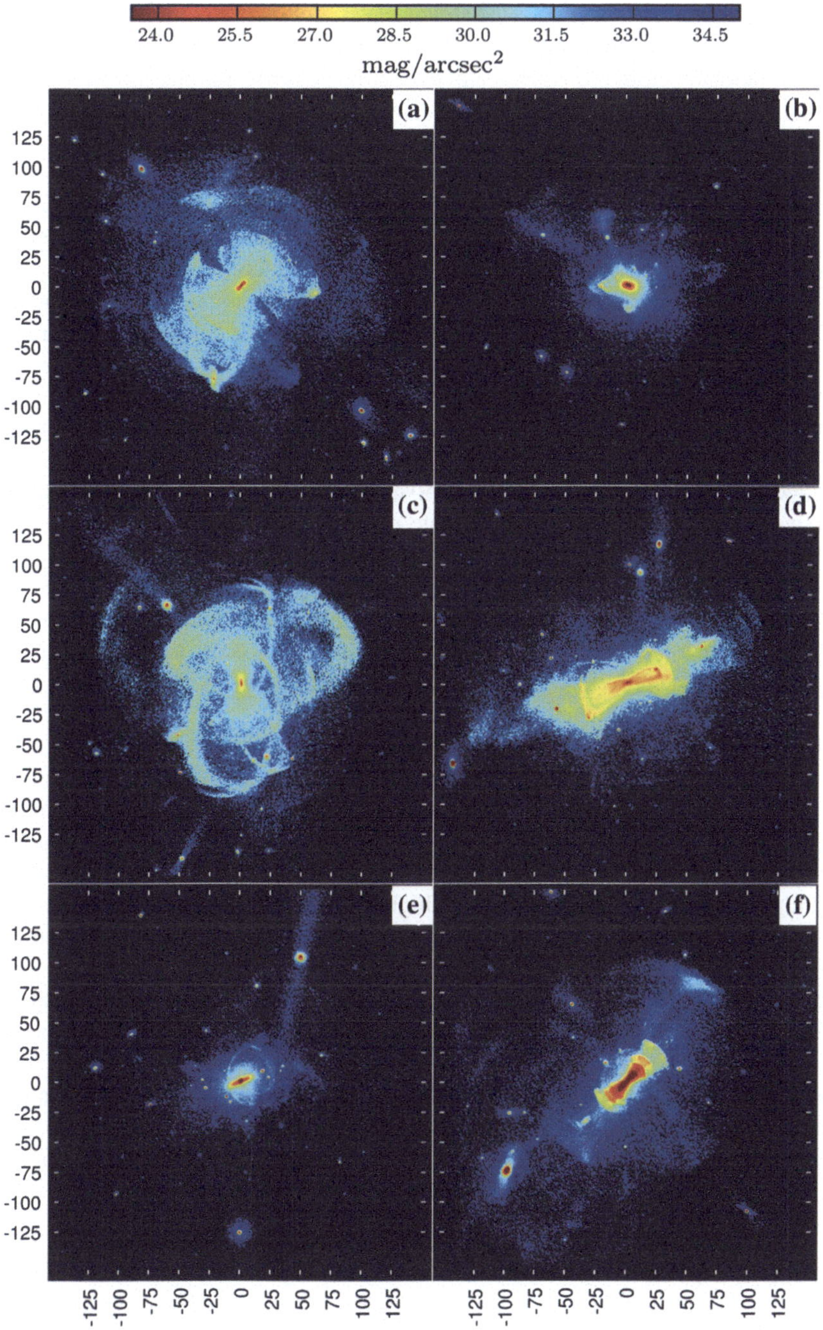


**Fig. 5** Surface brightness mocks images of Milky Way mass galaxies from the particle-tagging simulations of Milky Way galaxies from Cooper et al. (2010), MNRAS, 406, 74

simulations, the bulk of the accreted stellar material would have settled at the center of the galaxy. Initial tests using stacks of mock images from large cosmological hydrodynamical simulations indicate that it is possible to measure the average amount of total accreted stellar material by measuring the stellar mass beyond a certain physical radius where the accreted stellar material begins to dominate. We are presently measuring the average from the real SDSS stacks.

Armed with these techniques of stacking galaxy images together and estimating the accreted stellar fraction of the galaxy stack, we will be in a position to estimate how the average accreted stellar fraction varies as a function of galaxy stellar mass and type, and thus compare them with simulations. Preliminary analysis indicates that the average fraction of accreted stellar material in a galaxy is an increasing function of their stellar mass. In general, smaller galaxies have lower average accreted fractions, while larger galaxies have higher average accreted fractions. Moreover, the average fraction of accreted stellar material is much higher in elliptical galaxies than disk galaxies. Disk galaxies have very small average accreted stellar fractions from about 1–30% for the largest disk galaxies. Elliptical galaxies, on the other hand, have average accreted stellar fractions which increase from 20% for the smallest galaxies to as much as 70–80% for the larger ones. Many of these large elliptical galaxies sit at the center of large clusters of galaxies.

Furthermore, it will be possible to explore from these stacks how the average accreted stellar fractions vary with other galaxy parameters for a given stellar mass. This become especially interesting for Milky-Way size galaxies, where we expect the maximum spread in their accretion histories.

This will be the first time that the average accreted stellar mass fractions of galaxies have been measured over such a large range in galaxy stellar mass and type, and provides important observational constraints on the formation processes of galaxies. If the observational results are consistent with our current theoretical predictions, it will validate our current understanding of how galaxies form. If galaxies grow through merging, then larger galaxies will have a higher accreted stellar fractions than smaller galaxies. Since elliptical galaxies have undergone much more mergers than disk galaxies, their average accreted stellar fractions are also significantly higher.

The results of these studies complement information obtained by studying individual resolved stars in the stellar halo of our Milky Way galaxy and of other neighboring galaxies, as well as deep integrated light studies of the stellar haloes of nearby individual galaxies. Both these efforts are time consuming and are limited in the number of galaxies they can study. However, they have given us important insights into the stellar haloes of individual systems and the large variations which exist between them.

## Future Outlook

Given the history of the Vatican Observatory in projects like the Carte du Ciel, the VATT is well posed to undertake a survey of the stellar haloes of nearby galaxies. The VATT with its low focal ratio is especially suited for low surface brightness observations. This will complement the past research of Fr. Jose Funes in the star formation of nearby disk galaxies. Studying the Intra Cluster Light (ICL) of galaxy clusters would be a natural extension of studying the stellar haloes of galaxies due to their similar processes involved in their formation. This will extend the work of Fr. Omizzolo who studies galaxy clusters and is part of the WINGS (WIde-field Nearby Galaxy-cluster Survey) team.

Furthermore, in the age of large surveys like the LSST (Large Synoptic Survey Telescope), HSC-SSP (Hyper Suprime-Camera Subaru Strategic Program) and the DES (Dark Energy Survey), we will be able to probe individual galaxies to a greater depth and further out into their stellar haloes. Ongoing and future spectroscopic surveys of galaxies targeting the stellar halo of galaxies will allow us to understand the different type of the stars that make up the stellar halo and the environments in which these stars were born. These studies will help us to further refine our understanding of the various processes responsible for galaxy formation.

On the other hand, much more time and effort needs to be put into model the stellar haloes of galaxies. With the wealth of information which will come in with future surveys and missions, we will be need to understand how to reduce the complex multi-dimensional space of the accretion history of a galaxy in terms of the properties of its most important progenitors.

## Conclusion

Stellar haloes offer a unique way of probing the accretion history of a galaxy. In this paper, we highlight the important technique of stacking similar galaxies together from large imaging surveys to study the average properties of the stellar haloes of galaxies as a function of galaxy stellar mass and galaxy type. This allows us to quantify how much extra light (and stellar mass) is embedded in the galaxy's outer parts. Additionally it allows us to test our current understanding of galaxy formation and evolution, that galaxies grow hierarchically in the Universe.

## References

Cooper, A. P., Cole, S., Frenk, C. S., et al. (2010). Galactic Stellar haloes in the CDM model. *MNRAS, 406,* 744.

D'Souza, R., Kauffman, G., Wang, J., & Vegetti, S. (2014). Parametrizing the Stellar Halo of a galaxy. *MNRAS, 443,* 1433.

D'Souza, R., Vegetti, S., & Kauffman, G. (2015). The massive end of the stellar mass function. *MNRAS, 454,* 4027.

Martínez-Delgado, D., Gabany, J., Crawford, K. (2010). Stellar tridal streams in spiral galaxies of the local volume. *The Astronomical Journal, 140,* 962.

# Part IV
# Cosmology

# Aspects of Duality in Cosmology

**Gabriele Gionti, S.J.**

**Abstract** In the first part of this article, given the intent to stay at a popular level, it has been introduced and explained briefly basic concepts of Einstein's General Relativity, Dark Matter, Dark Energy, String Theory Quantum Gravity and Extended Theories of Gravity. The core of this research is based on selecting a class of theories of gravity, which exhibits scale factor duality transformations. The starting point of this theory is the effective theory of gravity derived from Bosonic String Theory which is called tree level effective theory of gravity. It is shown that this theory can be cast in a class of theories of gravity (modified theories of Einstein's General Relativity). It is imposed that Friedman-Lemaitre-Roberton-Walker (from now on FLRW) metric be solution of this class of theories, and, using the Noether symmetry approach, it is found that the cosmological model has scale factor duality like the Pre-Big Bang cosmology of Gasperini and Veneziano.

**Keywords** Modified theories of gravity · String theory · Scale factor duality Conservation laws

**PACS numbers** 95.35.+d · 95.36.+x · 04.50.kd · 11.25.-w

---

G. Gionti, S.J. (✉)
Specola Vaticana, 00120 Vatican City, Holy See (Vatican City State)
e-mail: ggionti@specola.va

G. Gionti, S.J.
Vatican Observatory Research Group, Steward Observatory, The University of Arizona, 933 North Cherry Ave., Tucson, AZ 85721, USA

G. Gionti, S.J.
Laboratori Nazionali Di Frascati, Istituto Nazionale Di Fisica Nucleare (INFN), Via E. Fermi 40, 00044 Frascati, Italy

 
G. Gionti, S.J. and J.-B. Kikwaya Eluo, S.J. (eds.), *The Vatican Observatory, Castel Gandolfo: 80th Anniversary Celebration*, Astrophysics and Space Science Proceedings 51, https://doi.org/10.1007/978-3-319-67205-2_9

## Introduction

This essay starts with a brief description of what is General Relativity. The basic concepts of General Relativity, the concept of Space-Time, the principles of equivalence and general covariance are introduced and discussed. It is also emphasized that Einstein's General Relativity was born by a fundamental question of how one can formulate a theory of gravitation, which is not "an action at distance" but a field theory on the example of electromagnetism. Then, in Einstein's General Relativity, Space-Time becomes a physical entity, in the sense that it is deformed by the mass-energy of bodies.

The theory of Einstein's General Relativity has equations of evolution in which the quantity to determinate is the topology and the metric structure of Space-Time, which, in simple words, is the shape of the Space and Time in which one lives (for example the spatial shape of our Universe is the surface of a four dimensional sphere and the time is a semi-line, a line with an origin). Said in other words, solutions of Einstein's Equations are cosmological models of the Universe like the FLRW cosmological model (which is isotropic and homogeneous in the spatial dimensions). FLRW cosmology is called the standard model of cosmology but, in order to fit with experimental data, needs some extra ingredients which are Dark Matter and Dark Energy that together FLRW cosmology is know as $\Lambda$CDM model. Here $\Lambda$ is the cosmological constant, which is introduced in the Einstein's Equations in order to include Dark Energy, and CDM stands for Cold Dark Matter. In fact, after many years of investigation it has been found that the dark matter known in the universe is at non relativistic energy (therefore it is "cold").

The main reason for introducing Dark Matter is due to the fact that, as it will be explained in the following, it is impossible to explain the distribution of the radial velocities of spiral galaxies without introducing the hypothesis of an existence of some kind of not visible matter. On the other side, the existence of Dark Energy is needed to explain the recent observed acceleration of the universe.

In this context, it is introduced the theory of extended theories of gravity (or alternative theories of gravity or f(R) theories), which are a generalization of Einstein's General Relativity and one of their aims is also to fit the recent cosmological data. In this respect, it is also mentioned the tests that the Planck collaboration has done to verify the validity of alterative theories of gravity.

The core of this work remains the effort of showing that the tree level effective theory of gravity derived from bosonic string theory can be written as a class of $f(R)$ theories of gravity. For this purpose, it has been explained what is String Theory and, in general, what is a Quantum Theory of Gravity. Then, it is highlighted that Bosonic String Theory, at low energy, when the gravitational interaction dominates, has a behavior like a Brans-Dicke gravity with a scalar field that is called dilaton. It is shown that this type of theory can be cast, using Weyl's transformations and a hypothesis on the relation between the Weyl's mode and dilaton field, in a class of $f(R)$ theories.

In this context, it has been stressed the issue of Duality in String Theory. In general, duality is, in string theory, a symmetry of the solutions of the equations of motion. Two different theories (two theories with different Lagrangian functions) exhibit the same solutions. In particular, duality symmetry on the tree level effective theory of gravity, which derives from string theory, is a symmetry both of the theory (the Lagrangian function) and of the equations of motions and this generates cosmological models, which are called pre-Big Bang models or Gasperini-Veneziano models.

In casting a tree level effective theory of gravity derived from bosonic string theory into a $f(R)$ theory, it has been used the Noether symmetry approach. This is a method, which fixes appropriate mathematical relations under which a $f(R)$ theory of gravity has a Noether symmetry, which simplifies the problem. Substituting this conditions into the class of $f(R)$ theories of gravity, it has been shown that it is possible to have that the class of $f(R)$ theories gets a duality symmetry, in the sense that there exists a duality symmetry of the type a(t) $\mapsto$ 1/a(t). It is straightforward to recognize that this symmetry is a scale factor a(t) inversion as in the Gasperini-Veneziano model.

## General Relativity

It is very nice coincidence that the 80th Vatican Observatory Anniversary symposium happens in the same year of the centennial jubilee of Einstein's General Relativity. Before Einstein's General Relativity, the gravitational force was described by Newton's Universal Gravitational Law. The universal gravitational law was "an action at distance", in the sense that if there were two massive bodies, the perturbation of the position of one of the two bodies will affect the second body "immediately", independently by their relative distance. This implies that the signal of this perturbation will propagate with an infinite velocity. The discovery of the electromagnetic field and the fact that its perturbations propagate with a finite velocity, that's the light velocity, raised two questions: could the gravitational force be described by a field theory? Does the velocity of propagation of the gravitational field be the velocity of light?

The answer to this question was found by Einstein's General Theory of Relativity. He found a theory of the gravitational field that is a field theory. In Einstein's General Relativity (from now on G.R.) the perturbations of the Gravitational Field propagates with the velocity of light. G.R. is an extension of Special Relativity. Special Relativity is a theory in which all the physical phenomena are described by physical laws, which are invariant in form (the same mathematical expressions) if one passes from one reference frame to another reference frame that moves with a constant velocity respect to the first one (this class of reference frames are called "inertial reference frames"). In this class of reference frames, the velocity of light is, by assumption, the same in every reference frame. In General Relativity the physical phenomena are described by physical laws, which are invariant ("covariant" as said in the jargon of G.R., which means that their mathematical expressions are the same) for transformations from one reference frame to another reference frame, which has an arbitrary velocity respect to the first one (non inertial reference frames). The velocity of light is always the same also in non-inertial reference frames.

Summarizing the previous consideration, one can finally state that Einstein's General Relativity is a field theory of gravity, which is based on two principles: the principle of mass equivalence and the principle of covariance. The principle of mass equivalence says that gravitational mass, the property for which two bodies attract each other by the gravitational force, is the same quantity for which a body oppose resistance to the motion, property that is in called "inertia". That is gravitational mass and inertial mass are the same thing and have the same numerical value.

The second principle of G.R. is the covariance principle. This principle says that the laws of physics are the same (this property is called covariance and means that the laws of physics have the same mathematical expression) in every arbitrary reference frame.

Space-Time becomes a physical entity, that is a dynamical quantity which is modified by the presence of massive bodies as well as (since the equivalence between mass and energy) the presence of energy. This is one of the greatest difference between General Relativity and any other field theory in physics. In fact, in any other field theory, Space-Time is like a box in which all the physical phenomena happens without any influence on it. On the contrary massive bodies and energy, in General Relativity, modifies Space-Time by generating a curvature on it, located around them. Free falling bodies move along the analogous of straight lines on the Space-Time surface, which are called geodedic lines. A geodedic line, by definition, is the line which makes stationary the functional of the distance between two points on the Space-Time surface.

In this way, any perturbation of mass-energy propagates through Space-Time, which acts like a medium. This propagation happens always, by construction, at the velocity of light. Therefore General Relativity is a field theory of the Gravitational Field.

## *Einstein's Equations*

The Gravitational Field, in analogy with the Electromagnetic Field, is described by Field Equations. As it is usually done in Field Theory, Field Equations are derived by a variational principle, which, for Einstein's theory of Gravitation, consists in finding the stationary points of the action functional $S[M, g]$. Here $S[M, g]$ is a functional on a Lorentian Manifold $(M, g), M$ being a Manifold and $g$ a Lorentian metric on it,

$$S[M, g] = \int_M R\sqrt{-g}d^4x \tag{2.1}$$

$R$ being the trace of the Ricci's tensor, which is a function of g. The variation of $S$ gets the Einstein's Equation in the vacuum (without matter)

$$R_{\mu\nu} - \frac{1}{2}g_{\mu\nu}R = 0 \tag{2.2}$$

These field equations are six independent equations and the metric tensor $g$ is the quantity to determinate. The metric tensor is a rank two (roughly speaking a matrix) covariant tensor, which is a function of space-time. Once it is determined the metric tensor, the topology of Space-Time can be determined via the definition of open and closed sets through the metric structure.

Einstein's equations are six independent equations, four of which are first order equations in the time derivatives. Therefore, they are not equations which give a dynamical evolution of the system (Dynamical evolution is given by equations that have the second time derivative of the quantity to determinate). They are constraint equations, that is limitations on the possible on the Cauchy's data (initial data) of the system.

The number of equations which generate the evolution of the dynamical system is, finally, two and it is the reason for which the graviton particle, that is also the field associated at fundamental level to the gravitational interaction, has two degrees of freedom.

## Dark Matter and Dark Energy

In recent years, astronomers have noticed that, studying the distribution of the radial velocity of matter as function of the distance from the center of galaxies, the mass observed does not explain the velocity distribution. The measurements of the radial velocity distribution can be explained well if one makes the hypothesis that there exists more matter than the observed one. This extra matter has been called Dark Matter, because it is not visible in the optical range (Peebles 1993). This matter does not interact with radiation and it does not loose sufficiently kinetic energy to relax into the disk of galaxies, as it does baryonic matter (matter which contains nuclear particles, leptons and neutrons). This implies that this sort of matter is electrically neutral and it has been found that its velocity is far from being relativistic, therefore it is called *Cold*-Dark Matter.

The study of a double galaxy cluster 1E0657-558 (the "bullet cluster") has confirmed the existence of cold Dark Matter, which has only gravitational interaction with itself and with baryonic matter (Weinberg 2008). The amount of Dark Matter in the universe amounts at 26.8% of the total matter according to the recent measurements of the Planck satellite (2016a, b). Dark Matter behaves, gravitationally, as completely ordinary matter; for example, it causes gravitational lensing (Weinberg 2008).

Soon after the discovery of General Relativity, Friedmann-Lemaitre-Robertson-and Walker (FLRW) found a homogeneous and isotropic solution of Einstein's General Relativity which implied that the Universe was expanding. Einstein did not believe that this solution has a physical meaning, therefore introduced, in his equations, a so called "Cosmological constant", which kept the universe stationary (not expanding). Later, in 1929, it was discovered that all spectra of the galaxies exhibit a "Red Shift" of their absorption lines, which was a proof that the Universe was expanding. Then Einstein said that he did the worst mistake of his life (Wald 1984).

In the early eighties of the last century, this mechanism of the cosmological constant was reconsidered for explaining an early period of the Universe in which it was thought to have expanded very fast, with an exponential rate, such that it could make sense to consider that all the different parts of the Cosmic Microwave Background Radiation were originally at thermal equilibrium. Cosmic Microwave Background Radiation is the "first light" from the Universe that was released 380 million of years after the Big Bang (Peebles 1993). Before, the Universe was too dense and every photon emitted was reabsorbed. Without invoking the mechanism of "Cosmological Inflation", FLRW metric solution of Einstein's Equations has a "particle Horizon", which means that not all the different regions of the Universe could, originally, be causally connected and then at thermal equilibrium. Cosmological Inflation, with an exponential expansion, implies an acceleration opposite to the gravitational attraction. This phenomenon is believed to be caused by the quantum vacuum of a field called the "Inflaton". This vacuum generates a huge contribution in the Einstein's General Relativity Equations driven by the cosmological constant, the same that before Einstein introduced and after withdraw when the galaxies' redshift and then the expansion of the universe was discovered (Wald 1984). This produces, in the equations of motions a negative pressure which gives reason to the emergence of a sort of anti-gravity force (Weinberg 2008). This mechanism generates an exponential expansion, which causally connects all regions of the early Universe that in standard FLRW metric are disconnected.

In 1998 observations of Type Ia supernovae showed that the Universe is expanding accelerating (Weinberg 2008). This result implies, as in the case of Cosmic Inflation, that there exists a kind of force, which is opposite to Gravity. The mechanism known, up to now, to explain this antigravity force is "Dark Energy". Dark Energy is believed to be the vacuum point (zero point) energy of fundamental quantum fields. In fact, the zero-point energy of the harmonic oscillator, which is, basically, the system on which any quantum filed theory is based, is not zero but has a finite value. The sum of all these vacuum modes, for each fundamental interaction, generates an energy ("Dark Energy" because one does not see it), which is present in the Einstein's Equations via the cosmological constant term. The last cosmological measures, obtained through the Planck satellite, seem to point out that the best theory capable to explain the cosmological measures is the $\Lambda$CDM model, that is the Einstein's theory of General Relativity with Cold Dark Matter and Dark Energy. In the ΛCDM model, the amount of Dark Energy is 68,3% of the total Universe mass (2016b) and Einstein's General Relativity still remains the theory that explains better the cosmological data than other theories of gravity.

## Extended Theories of Gravity

Extended theories of Gravity are "generalization" of Einstein's General Relativity. The principles, on which Extended theories of Gravity are based on, are the same of Einstein's General Relativity. The only, main, difference concerns the dynamics in the sense that their "action" and the relative equations of motions are, in general, respectively, generic functions of the trace of the Ricci tensor and equations of motion with higher order, more than two, derivatives of the metric tensor.

Modifications to Einstein's General Relativity has been proposed since 1970 (see Sotiriou and Faraoni 2010) but mainly with the beginnings of the theory of cosmological inflation (Starobinsky 1980) it was clear that modifications of Einstein's General Relativity were important to understand the early period of our Universe. One of the "dreams" of extended theories of gravity has been to fit the experimental data, as showed in (Capozziello and De Laurentis 2012), with extended theories of gravity without introducing dark matter. But these models have too many parameters and appear quite phenomenological (see Capozziello and De Laurentis 2012). Recent work of the Planck Collaboration (2016b) have tried to study, on the base of recent data collected by the Planck satellite, how well the experimental data fit with modified theories of gravity.

Extended theories of gravity are important also since, in many cases, they are the low energy limit of fundamental theories of Quantum Gravity. This, in simpler words, means that Extended theories of gravity can be considered, when the energies are not so high as in fundamental theories, a good approximation of the physical laws that regulates the quantum world in the very early times of our Universe. For example, String Theory, which is an attempt to find the quantum mechanical behavior of the gravitational field, in the limit of lower energies, when the gravitational interaction is supposed dominating over the other interactions, behaves as an Extended theory of gravity (Tseytlin 1991).

One serious problem of extended theories of gravity is their quantization. In fact, once quantized, extended theories of gravity suffer lack of unitarity (Capozziello and De Laurentis 2011), which means, at quantum mechanical level, in analogy with classical mechanics, that if one changes the reference frame, Physics is affected by this change in the sense that the physical laws in the new reference frame appear different. This is one of the reasons for which people say that they are phenomenological theories and not fundamental, that is to say that they are "ad hoc" theories that works at same scales of energies but not at higher energies.

An important aspect of alternative theories of gravity is the result (see De Felice and Tsujikawa 2010) that $f(R)$ theories are equivalent to Brans-Dicke theory. This is a theory which, as concern its Lagrangian function, has the following characteristic: the trace of the Ricci tensor is multiplied with a scalar field; this scalar field has a potential and a kinetic term. It can be also shown that if one implements Weyl transformations on the metric tensor and choose a particular Weyl transformation, which is field dependent, (De Felice and Tsujikawa 2010) a $f(R)$ theory can be shown to be equivalent to Einstein's Gravity with a potential and a kinetic term related to the scalar field.

## String Theory

String Theory is one of the proposal for a Quantum Theory of Gravity (Quantum Gravity). As it is quite well known, Einstein's equations have a singularity around the point t = 0. This means that the physics, one knows, does not work around this initial point. The Physics that should describe our Universe around the point t = 0 is called Quantum Gravity. A theory of Quantum Gravity is not known yet, there are many attempts to define a theory of Quantum Gravity, but no one can say that there exists a definitive formulation of Quantum Gravity (see Rovelli 2004). There exist many proposals for a Quantum Theory of Gravity. They can be divided into two main principal lines. A line that makes the assumption that the quantization of the gravitational field has to be performed through the quantization of Einstein's General Relativity, therefore without including all other fundamental interactions, and a line that promotes that the quantization of the gravitational field through the unification of all fundamental forces at the energy for which (it is called Planck Energy) the laws of the gravitational field should behave according to Quantum Mechanics. As it is well known, Quantum Mechanics describes the laws of physics at microscopic level, that is at atomic and subatomic levels. Apart from gravity, the others fundamental interactions are: the Electromagnetic, the Weak Nuclear and the Strong Nuclear interaction. The Electromagnetic force regards the interaction among particles that have an electric charge. The Weak Nuclear force regards subatomic particles, which interact "weakly" (respect to the strong interaction), whose interactions are described by particle mediators which are massive. The Strong Nuclear force deals with particles that interact "strongly" and are called Adrons.

String Theory is a theory born at the end of the 60s of the past century and it is considered a theory of Quantum Gravity. It is based on the "principle" that all fundamental forces in nature, the Electromagnetic Force, the Strong Nuclear Force, the Weak Nuclear Force and the Gravitational Force can be unified through a unique object that is called String. Particles are not, in this approach, the last building blocks of matter, but they are considered composite objects made out of Strings. The quantum oscillations of Strings will generate particles in the way one knows. Strings are one dimensional objects that could be open (Open Strings) and closed (Closed Strings). First excited level of Strings shows that their spectra contain particles of all the fundamental interactions. In this way the String, having in its first excited states all different particles of different interactions, realizes the unification of all the fundamental forces (Green et al. 1986) one knows in nature.

Strings, in order to make sense mathematically, have to be embedded into a background space, which, for bosonic String Theory, is 26-dimensional. Bosonic String Theory is a theory that deals only with particles with integer spin (bosons). In fact, it happens that bosonic string theory has an "anomaly", which is called conformal anomaly. Anomaly means that a classical symmetry is not conserved at quantum level. Conformal Anomaly is a behaviour of the String Theory for which the classical theory is invariant under some specific transformations of the metric tensor that are called "conformal transformation", while the correspondent quantum theory results not invariant under conformal transformations. Therefore, in order to make the theory conformal invariant also at quantum level, one finds that has to embed the string into a background space of 26 dimensions. People have considered also Supersymmetric String Theory, which is String Theory in which there is a further symmetry, which is called supersymmetry. This symmetry implies that it is possible to exchange particles with integer spin (Boson) with particle with semi-integer spin (Fermion) or, said in a different way, for ever integer spin particle there is a semi-integer spin partner particle. Supersymmetric Strings need to be embedded in a 10-dimensional space in order to avoid the conformal anomaly.

At one loop, that means at the first order of approximation, at quantum level, in the case of Bosonic String Theory, the fact that the conformal symmetry has to be preserved implies that the trace of the Ricci's tensor plus the square of the first (covariant) derivatives of the scalar field, plus the D'Alambertian of the scalar field, plus the square of the derivatives of a field which is called B-field, plus a constant, in bosonic string theory in non critical dimensions (= different from 26 dimensions), have to be zero. These look similar to Einstein's General Relativity equations. Therefore this result is considered an argument in favor of the fact that the Gravitational interaction is contained in Bosonic String Theory (Green et al. 1986).

One important feature of Bosonic String Theory is that it exibits a symmetry which is called "Duality". If the Lagrangian function of Bosonic String Theory has a particular symmetry, which means that there exists a vector field, an infinitesimal transformation, for which the Lagrangian function results to be invariant, then it is possible to define "dual picture", dual fields that determine a new Lagrangian function, whose equations of motion have the same solutions of the starting Lagrangian. This is an example of Duality and in particular is called Busher's Duality. In general, a scalar field called the "dilaton" is included in the mathematical formulation of Bosonic String Theory. The mathematical consequence of the preservation of the conformal invariance, as already said, implies that Einstein's equations with cosmological constant (in the case of non-critical string, that is the background Space-Time, in which the String is embedded, is not 26 dimensional) and a scalar field is satisfied. It can be eaisly shown that these equations derive from an action, that is the Einstein-Hilbert 's action plus a cosmological constant, plus a kinetic part of the scalar field (the dilaton) everything multiplied by the exponential of the dilaton field, which is a Brans-Dicke theory like (see for all details Capozziello et al. 2016; Gabriele Gionti 2016). Generally speaking, the Lagrangian function could have more than one symmetry. If one has a theory which has several symmetries, in particular n independent symmetries which, in mathematical term, is said n independent and commuting isometries, then it is possible (see de la Ossa and Quevedo 1993) to define a dual theory such that the original theory and its dual, put together, are invariant under particular sets of transformations, which are called the group O(n, n).

Gasperini and Veneziano (2003) have studied into great details the cosmological models of the gravitational field which derives from bosonic string theory. As it has been extensively said, this is a Brans-Dicke like gravitational field with a kinetic term in the scalar field (dilaton), with the presence of the derivatives of a "B-field" (which is usually put to zero) and a cosmological constant in the case of non critical dimensions. This theory is manifestly O(n, n) invariant (Meissner and Veneziano 1991), therefore one finds that, imposing that the theory has flat FLRW metric as solution of the equations of motions, there exists Duality symmetry transformations of the action and of the equations of motions (Tseytlin and Vafa 1992). This theory is at the base of the Pre-Big Bang cosmology scenario in which there exists a phase of our universe before the time t = 0.

## From String Theory to Modified Gravity

Suppose that in Bosonic String theory the gravitational interaction is dominating over the other interactions. In this case one knows, of course, that the graviton mode represents the main interaction. In this hypothesis, in the low energy limit at tree level Bosonic String theory becomes a Brans-Dicke-like gravity theory coupled to the dilaton field and with a B-field (a field which emerges in the low energy limit of Bosonic String Theory) and a cosmological constant (in the case in which one works in non critical dimensions, n). One question is if this theory can be cast in a $f(R)$ theory of gravity. One of the tracks followed has been to find a Weyl transformation on the metric tensor in such a way that the action of gravity, which derives from string theory, can be re-written as an alternative theory of gravity, a $f(R)$ theory (see Capozziello et al. 2016; Gabriele Gionti 2016). A Weyl transformation is a transformation of the metric tensor, in which the metric tensor is equivalent to another metric tensor, in a different reference frame, multiplied by a generic scalar function (the Weyl's factor).

The equivalence, in practice, is get by equating the integrands of the actions of the string based effective theory of gravity (with $B = 0$) and the $f(R)$ theory. Using some results from the equations of motion and imposing that the Weyl's factor is linked to the dilaton field, it is possible to pin down a functional form for a class of $f(R)$ functions, which derive from bosonic string theory.

Next step of this research is to study the cosmological consequences of this particular classes of $f(R)$ functions. For this reason, one consider a FLRW metric in the flat case. FLRW universe is like, in two dimensions, the surface of a sphere which is expanding and its radius measures the time, since the Big Bang, of the universe. In particular, one, for simplifying the problem, restricts to the flat case in which the space is flat and the distance between points in the space increases with time.

Imposing that this metric be solution of these $f(R)$ theories, one obtains a Lagrangian function, which depends also from the scale factor of the Universe.

Given this particular classes of $f(R)$ functions, following a general method, one implements the FLRW metric solution introducing a new action functional in which there is the $f(R)$ function plus a Lagrange multiplier which constrains the trace of the Ricci curvature to be equal to the Ricci curvature evaluated on FLRW metric.

At this point, in order to make the theory mathematically simpler, one uses a technique quite useful in many cases of study for $f(R)$ theories of gravity, which is called Noether symmetry approach. Noether symmetry approach is a procedure, which studies the general properties that a $f(R)$ theory of gravity has in order that its Lagrangian function has a symmetry. In general, a physical system may have some type of symmetries, like the translational symmetry, the rotational symmetry etc. These kinds of symmetries become groups of transformations. They are symmetries of the original Lagrangian function and of the solutions of its related equations of motion. In fact, these transformations map solutions of these equations of motion into new solutions of the same equations of motion.

Once one imposes that a $f(R)$ function has to have Noether symmetries, this translates into mathematical conditions on the $f(R)$ functional form itself. In particular, considering that the class of $f(R)$ functions derived from bosonic string theory has to satisfy these Noether symmetries conditions, one gets a specific functional form for this class.

Finally, one can renormalize the dilaton field by adding a function of the scale factor a(t) of the universe without modifying the equations of motion. As result, one gets a Lagrangian function which is manifestly duality invariant under the transformation on the scale factor of the universe, $a(t) \mapsto 1/a(\pm t)$, which is the same transformation of the Gasperini-Veneziano pre-Big Bang cosmological models. In this way, from any cosmological solution valid for the scale factor of the universe $a(t)$ one can derive duality symmetric solutions with scale factor of the universe $1/a(\pm t)$. As it is quite clear from the duality transformations, each time one has a cosmological solution valid for positive values of time, it is possible to get a solution for negative values of time as well. Therefore a cosmological model with a starting Big-Bang will have a specular solution that develops before the Big-Bang time and ends in the neighbor of it.

## Conclusions

In this essay, it has been highlighted that extended theories of gravity are more than "ad hoc" phenomenological theories since, as in same cases treated above, they derive directly from low energy limit of fundamental theories of quantum gravity like string theory.

Potentially, Extended theories of gravity could shed light in the enigma of understanding the problems of Dark Matter and Dark Energy. This is especially true for those extended theories of gravity that are approximation of Quantum Theory of Gravity.

The very fact, one has highlighted, that there exist classes of $f(R)$ theories of gravity which are low energy limit of Quantum Gravity Theories is quite important. From the cosmology of these theories it could develop a cosmography, whose agreement with the observational data could discriminate among the different theories of quantum gravity.

Furthermore, the fact that one has derived the duality symmetry by using Noether symmetry approach suggests to explore the link between the Noether symmetries and duality transformations. The fact that the effective theory of gravity derived from string theory has a manifestly $O(d, d)$ invariance, is questioning how this is linked to the Noether symmetries which have been used to make the class of $f(R)$ Lagrangians manifestly duality invariant.

## References

Peebles, P. (1993). *Principles of physical cosmology*. Princeton (NJ): Princeton Series in Physics.

Weinberg, S. (2008). *Cosmology*. Oxford (U.K.): Oxford University Press.

Planck Collaboration. (2016a). Planck 2015 results XIII. Cosmological Parameters. *Astronomy & Astrophysics, 594*, A13. [arXiv:1502.01589].

Planck Collaboration. (2016b). Planck 2015 results XIV. Dark Energy and Modified Gravity. *Astronomy & Astrophysics, 594*, A14. [arXiv:1502.01590].

Capozziello, S., & De Laurentis, M. (2011). Extended theories of gravity. *Physics Report, 509* (4–5), 167–321. [arXiv:1108.6266].

Capozziello, S., & De Laurentis, M. (2012). The dark matter problem from $f(R)$ gravity point of view. *Annalen der physik, 10, 524* (9–10), 545–578.

Capozziello, S., Gionti, G., S.J., & Vernieri, D. (2016). String duality transformation in $f(R)$ gravity from Noether symmetry approach. *Journal of Cosmology and Astroparticle and Physics, 1601*, 015. [arXiv:1508.00441].

De Felice, A., & Tsujikawa, S. (2010). $f(R)$ theories. *Living Review in Relativity, 13*, 3. [arXiv:1002.4928].

de la Ossa, X., & Quevedo, F. (1993). Duality symmetries from non-Abelian isometries in string theory. *Nuclear Physics B, 403*, 377–394. [arXiv:9210021].

Gasperini, M., & Veneziano, G. (2003). The Pre-Big Bang scenario in string cosmology. *Physics Reports, 373*, 1–212. [arXiv:0207130].

Gionti, G., S.J. (2016). *O(d, d)* duality transformations in F(R) theories of gravity. In *Proceeding of the XIV Marcel Grossmann meeting*, Rome, July 12–18, 2015. [arXiv:1608.05223].

Green, M., Schwarz, J., & Witten, E. (1986). *Superstring theory*. Cambridge (U.K.): Cambridge University Press.

Meissner, K., & Veneziano, G. (1991). Manifestly *O(d, d)* invariant approach to space-time dependent string vacua. *Modern Physics Letters A, 6*, 3397–3404. [arXiv:9110004].

Rovelli, C. (2004). *Quantum gravity*. Cambridge (U.K.): Cambridge University Press.

Sotiriou, T., & Faraoni, V. (2010). $f(R)$ theories of Gravity. *Review of Modern Physics, 82* (1), 451. [arXiv:0805.1726].

Starobinsky, A. (1980). A new type of isotropic cosmological models without singularities. *Physics Letter B, 91*, 99–102.

Tseytlin, A. (1991). Duality symmetric closed string theory and interacting chiral scalars. *Nuclear Physics B, B350,* 395.
Tseytlin, A., & Vafa, C. (1992). Elements of string cosmology. *Nuclear Physics B*, *372*, 443. [arXiv:9109048].
Wald, R. M. (1984). *General relativity.* Chicago: Chicago University Press.

# Creation of Matter in a Noncommutative Universe

**Tomasz Miller and Michael Heller**

**Abstract** The dark matter and dark energy problem, that is now dominating the research in cosmology, makes the question of the origin of mass-energy content of the universe more urgent than ever. There are two philosophies regarding this question: according to Mach's principle it is matter that generates geometry of space-time, and according to Wheeler's geometrodynamics some configurations of space-time geometry are to be interpreted as its material content. Neither of these philosophies has led to success. In the present paper, we show that there exists an algebraic generalisation of geometry that reconciles, in a sense, these two seemingly opposite standpoints. The geometry is constructed with the help of a noncommutative algebra of smooth functions on a groupoid and its derivations. The groupoid in question has a nice physical interpretation: it can be regarded as a space of Lorentz rotations. In this way, Lorentz symmetries are inherent to the generalised geometry of space-time. We define the action for this geometry and, by varying it, obtain generalised vacuum Einstein equations (for a simplified model). It turns out that these equations contain additional terms (with respect to the standard vacuum Einstein equations) which are naturally interpreted as the components of the energy-momentum tensor. Matter is thus created out of purely geometric degrees of freedom. We find two exact solutions (for even more simplified case). We argue that the creation of matter, being a global effect, makes the contrast between Mach and Wheeler philosophies ineffective.

T. Miller (✉)
Faculty of Mathematics and Information Science,
Warsaw University of Technology, ul. Koszykowa 75, 00-662 Warsaw, Poland
e-mail: T.Miller@mini.pw.edu.pl

T. Miller · M. Heller
Copernicus Center for Interdisciplinary Studies, ul. Szczepanska 1/5,
31-011 Cracow, Poland

M. Heller
Vatican Observatory, V-00120 Castel Gandolfo, Vatican City, Italy



G. Gionti, S.J. and J.-B. Kikwaya Eluo, S.J. (eds.), *The Vatican Observatory, Castel Gandolfo: 80th Anniversary Celebration*, Astrophysics and Space Science Proceedings 51, https://doi.org/10.1007/978-3-319-67205-2_10

## Introduction

The dark matter and dark energy problem, that is now dominating the research in cosmology, makes the question of the origin of mass-energy content of the universe more urgent than ever. This problem could hardly be solved without a deeper understanding of matter creation mechanisms in the universe. Einstein's equations are plagued with a dualism of geometry (left-hand side of the equations) and matter (their right-hand side). Einstein looked for a remedy against this difficulty in the idea called by him Mach's principle. It admits several nonequivalent formulations, some stronger of which claim that (local) geometry should totally be determined by the global distribution of matter throughout space-time. The idea could be encapsulated in the slogan "geometry out of matter" (Barbour and Pfister 1995). The opposite philosophy was propagated by John Archibald Wheeler who argued that the material sources that appear in Einstein's equations can entirely be reconstructed from a characteristic imprint they exert on the space-time geometry. Here we have a slogan "matter out of geometry". The program was known as "Wheeler's geometrodynamics" (Wheeler 1964; Wheeler 1968; Wheeler 1973). The fact that neither of these philosophies, attractive as they are, has led to the success suggests that a stumbling block lies in the very concept of geometry that is too rigid to accommodate for such a complex phenomenon as that of matter. In the present paper, we explore a generalisation of the standard differential geometry and, basing on it, construct cosmological models in which two above-mentioned philosophies not only seem to work, but also could, in a sense, be unified.

There are two ways of doing differential geometry: by using local coordinate systems, and algebraically in terms of smooth functions on a given space, the latter being independent of the choice of coordinates. Both these methods are equivalent, but the second is better suited for generalisations. Let us then consider a space-time manifold $M$ and the algebra $C^\infty(M)$ of smooth functions on $M$. The algebra $C^\infty(M)$ can, without losing its geometric properties, be replaced by a more general, not necessarily commutative, algebra on a more general space than the space-time manifold $M$. The more general space we use to this end is the so-called groupoid, denoted by $\Gamma$ (for details see Sect. 2). The choice of this space has a nice physical motivation. The groupoid concept is a generalisation of the group concept, and as such it generalizes the notion of symmetry. In our case, the elements of the groupoid $\Gamma$ can be regarded as Lorentz rotations. Therefore, $\Gamma$ can serve as a more natural environment for general relativity (being the space of its symmetries) than the usual "naked" space-time.

We then consider the algebra $C^\infty(\Gamma)$ of smooth (compactly supported) functions on a groupoid $\Gamma$. The consequence of these rather simple replacements is that space-time points acquire their internal structure, i.e. internal degrees of freedom not unlike in the Kaluza-Klein type models, with the difference that now internal degrees of freedom are really "internal", not created by adding new space dimensions.

The algebra $C^{\infty}(\Gamma)$ can easily be made into a $C^{*}$-algebra which assimilates the formalism to that employed in quantum mechanics. In fact, this approach has been used to construct a model unifying general relativity and quantum mechanics with a perspective to make an attempt at a quantum gravity theory (Heller et al. 1997; Heller and Sasin 1999; Heller et al. 2005; Heller et al. 2007). It goes without saying that truly fundamental mechanism of matter creation can hardly be imagined without the correctly working quantum gravity theory. This is why the present work can only be regarded as a preliminary step in this direction.

Models presented in this paper are nothing more than "toy models"; however, they seem worthwhile to be explored not only because they give such surprising results as far as matter generation is concerned, but also because they are preparing mathematical tools to deal with more realistic situations. The present work is based on our research paper (Heller et al. 2015).

The plan of our essay runs as follows. In Sect. 2, we briefly discuss the method of doing geometry on the groupoid $\Gamma$ in terms of the algebra $C^{\infty}(\Gamma)$ and its derivations. In Sect. 3, we construct such a geometry for a groupoid with a finite structure group. Then, in Sect. 4, we deduce generalised Einstein equations from the corresponding action principle, and show that they, when projected onto space-time, contain an additional term that can naturally be interpreted as a "matter source". We also find two explicit "Friedman-like" solutions for a simplified form of the metric. Finally, we append some concluding remarks.

## Space of Lorentz Symmetries and Its Geometry

A natural setting for relativity theory is a set of pairs of reference frames with a Lorentz transformation acting between them. This setting can be given strict mathematical form. Let $M$ be a space-time, and let us consider the bundle of reference frames (frame bundle, for short) on $M$, denoted by $(E, \pi_M)$ where $E$, called the total space of the bundle, is the set of all local reference frames on $M$, and $\pi_M : E \to M$ is a mapping projecting a frame $p \in E$ to its attachment point $x \in M$, $\pi_M(p) = x$. The set of all frames attached to $x$, $E_x = \pi_M^{-1}(x)$, is called the fiber at $x$. Let $G$ be a Lorentz group or one of its subgroups. It acts on $E$ along fibres, i.e. any two frames in the same fibre can be transformed into each other with the help of an element of $G$.[1] In this way, the Cartesian product has been constructed,

$$\Gamma = E \times G = \{\gamma = (p, g) : p \in E, g \in G\}.$$

It is clear that Lorentz transformations in a given fibre can, as elements of a group, be suitably composed and have inverses. The construction, just described, is

[1]This means that we consider only Lorentz rotations; translations, i.e. transformations between fibres, are excluded.

called transformation groupoid with $\Gamma$ as its groupoid space, $E$ its base space, and $G$ its structure group. This purely algebraic construction can be equipped with the smoothness structure; it is then called a smooth transformation groupoid.

Just as geometry of space-time $M$ can be done in terms of the algebra $C^{\infty}(M)$ of smooth functions on $M$, a generalised geometry of space-time $M$ can be done in terms of an algebra $\mathcal{A}$ on the groupoid $\Gamma = E \times G$. The usual algebra $C^{\infty}(\Gamma)$ of smooth (compactly supported) functions on $\Gamma$ (with pointwise multiplication, denoted by $\cdot$) would reproduce geometry of $\Gamma$ with nothing interesting for our program. To obtain an interesting generalisation we replace the commutative pointwise multiplication with a not necessarily commutative multiplication $*$. Its concrete form need not bother us here (for details see (Heller et al. 2015)). We thus consider the algebra $\mathcal{A} = (C^{\infty}(\Gamma), *)$ of smooth (compactly supported) functions on $\Gamma$. It is, in general, noncommutative which means that if $a, b \in \mathcal{A}$ then $a * b \neq b * a$. This seemingly innocuous modification leads to dramatic changes in geometry. We also impose on $\mathcal{A}$ suitable smoothness conditions.

What is the contact of the algebra $\mathcal{A}$ with the algebra $C^{\infty}(M)$ of smooth functions on space-time $M$? To answer this question we remind the concept of the center of a noncommutative algebra. The center of an algebra $\mathcal{A}$, denoted $\mathcal{Z}(\mathcal{A})$ is the set of all elements of $\mathcal{A}$ that commute with all elements of $\mathcal{A}$ (it is obvious that if an algebra is commutative, it coincides with its center). It can be shown that the center of our algebra $\mathcal{A}$ is isomorphic with $C^{\infty}(M)$. Therefore, if, in constructing geometry, we restrict the algebra $\mathcal{A}$ to its center $\mathcal{Z}(A)$, we obtain the geometry of space-time $M$.

The standard manifold geometry is encoded in its metric tensor. Mathematically, metric is a function taking two smooth vector fields as its arguments and returning a smooth function on this manifold. In the algebraic approach to geometry, the counterparts of vector fields are derivations of the corresponding algebra. Let us consider our algebra $\mathcal{A}$. Its derivation is a linear map $v : \mathcal{A} \to \mathcal{A}$ satisfying the well-known Leibniz rule

$$v(a * b) = v(a) * b + a * v(b)$$

for $a, b \in \mathcal{A}$. The set of all derivations of the algebra $\mathcal{A}$ is denoted by $\mathrm{Der}(\mathcal{A})$.

Taking the above into account, we assume the metric of the form $\mathcal{G} : V \times V \to \mathcal{Z}(\mathcal{A})$ where $V \subseteq \mathrm{Der}(\mathcal{A})$.[2] The pair $(\mathcal{A}, V)$ is called differential algebra. It serves us as the basic structure to construct an algebraic version of generalised geometry. We proceed in strict analogy to what is usually done when developing the standard differential geometry. We construct connection (with the help of the Koszul formula), curvature and all other magnitudes necessary to write down Einstein's field equations. In the following, when pursuing this program, we shall limit ourselves to a special case of a finite model.

[2] $V$ should have the $\mathcal{Z}(\mathcal{A})$-module structure as well as the Lie algebra structure.

## Geometry on the Groupoid Algebra with Finite Structure Group

Let us consider an $n$-element subgroup $G$ of the group of Lorentz rotations. Without loss of generality, we can assume that the frame bundle studied is trivial: $E = M \times G$, and therefore the considered groupoid is

$$\Gamma = M \times G \times G = \{(x, g_i, g_j) : x \in M, g_i, g_j \in G, i, j = 1, \ldots, n\}.$$

The noncommutative multiplication $*$ in the algebra $C^\infty(\Gamma)$ of smooth (compactly supported) functions on $\Gamma$ is given by

$$\forall a, b \in C^\infty(\Gamma) \quad (a * b)(x, g_k, g_l) := \sum_{m=1}^{n} a(x, g_k, g_m) b(x, g_k g_m, g_m^{-1} g_l).$$

It is much more convenient, however, to regard every function $a \in C^\infty(\Gamma)$ as an $n$-by-$n$ *matrix* $(a_{ij})$ with $C^\infty(M)$-valued entries, defined via

$$a_{ij}(x) := a(x, g_i, g_i^{-1} g_j).$$

In this way, the algebra studied becomes

$$\mathcal{A}_n := \mathbb{M}_n(C^\infty(M)) = C^\infty(M) \otimes \mathbb{M}_n(\mathbb{C})$$

and the operation $*$ becomes nothing but the standard matrix multiplication

$$\forall a, b \in \mathcal{A}_n \quad (a * b)_{ij}(x) := \sum_{k=1}^{n} a_{ik}(x) b_{kj}(x).$$

Let us note that algebras similar to $\mathcal{A}_n$ commonly appear in the exploration of possible applications of noncommutative geometry to physics, most notably in the context of the Noncommutative Standard Model of particle physics (see (Chamseddine et al. 2007; Connes 1994) for details on noncommutative geometry and the Noncommutative Standard Model or (Connes 2006; Jureit et al. 2007) for a more accessible review).

The center $\mathcal{Z}(\mathcal{A}_n)$ of this algebra consists of matrices of the form $f I_n$, where $f \in C^\infty(M)$ and $I_n$ denotes the $n$-by-$n$ identity matrix. It is, therefore, isomorphic to the algebra $C^\infty(M)$, just as the previous section anticipated. From now on we shall identify $\mathcal{Z}(\mathcal{A}_n)$ with $C^\infty(M)$.

We consider the full $C^\infty(M)$-module of derivations of the algebra $\mathcal{A}_n$, $V := \text{Der}\,\mathcal{A}_n$. It can be shown that $V$ decomposes into the direct sum of two its submodules:

- The submodule Hor $\mathcal{A}_n$ of *horizontal* derivations. Its elements are *liftings* of the smooth vector fields on the manifold $M$ onto $\mathcal{A}_n$. More explicitly, for any smooth vector field $X \in \mathrm{Der}\, C^\infty(M)$ one defines its lifting as a map $\bar{X} : \mathcal{A}_n \to \mathcal{A}_n$ acting entrywise, namely

$$\forall a \in \mathcal{A}_n \quad (\bar{X}a)_{ij} := Xa_{ij}.$$

- The submodule Inn $\mathcal{A}_n$ of *inner* derivations. By an *inner derivation* induced by an element $b \in \mathcal{A}_n$ one understands a map $\mathrm{ad}_b : \mathcal{A}_n \to \mathcal{A}_n$ defined as

$$\forall a \in \mathcal{A}_n \quad \mathrm{ad}_b a := [b, a] = b * a - a * b.$$

Embarking on the construction of the generalised geometry on the differential algebra $(\mathcal{A}_n, V)$, one begins with the *metric* $\mathcal{G} : V \times V \to \mathcal{Z}(\mathcal{A})$, upon which the Levi-Civita connection and the curvature tensors are subsequently defined. All these objects can be, and usually are, studied by physicists with the help of the so-called *abstract-index notation*. The idea behind this notation is to identify the abstract (usually tensorial) objects with the array of their components in a chosen basis of $V$. Multi-indexed expressions, obtained in this way, can then be effectively manipulated under the set of simple rules, among others including the Einstein summation convention (see Penrose and Rindler 1984, Chap. 2).

The natural choice for the (local) basis of Hor $\mathcal{A}_n$ is the lifting of the *coordinate basis* $\left(\frac{\overline{\partial}}{\partial x^\mu}\right)$ induced by some chart $x$. In the following, we shall denote the liftings of the coordinate vector fields $\frac{\overline{\partial}}{\partial x^\mu}$ simply by $\partial_\mu$, suppressing both the overline and the reference to the inducing chart and using also other *lowercase Greek letters* $\nu, \lambda, \sigma, \ldots$. We adopt the usual convention that these indices can assume values $0, 1, \ldots, m-1$, where $m = \dim M$.

On the other hand, one can show that any basis of Inn $\mathcal{A}_n$ contains exactly $n^2 - 1$ elements. In what follows, we shall not specify the basis concretely, but we shall denote its elements by $\partial_{\{A\}}$ (by analogy with the horizontal derivations), using also other *capital Latin letters in curly brackets* $\{B\}, \{C\}, \{D\}, \ldots$. These indices can assume values $\{1\}, \{2\}, \ldots, \{n^2-1\}$ (interpreted as the "inner degrees of freedom"). The use of curly brackets assures that we do not mix the values of indices of the two types.

Summarizing, the basis of $V$ contains two kinds of elements: the horizontal derivations and the inner derivations, and this is reflected in the two types of indices used: the lowercase Greek indices and the capital Latin indices in curly brackets. Additionally, it will be convenient to write $\partial_A$ for a generic derivation from the basis, using also other *capital Latin letters* $B, C, D, \ldots$. These indices can take all values assumed by the indices of the two types listed above.

Let us note here that the Einstein summation convention applies to each of the three types of indices separately.

The noncommutativity of $\mathcal{A}_n$ has a direct effect on the commutation relations between $\partial_A$'s. Even though $[\partial_\mu, \partial_\nu] = 0$, exactly as for the coordinate basis in the standard differential geometry, in general we have that $[\partial_A, \partial_B] \neq 0$. Mathematically speaking, some of the *structure constants* $\mathbf{c}^C_{AB}$, defined by the formula $[\partial_A, \partial_B] = \mathbf{c}^C_{AB}\partial_C$, are nonzero.

Having specified the basis, we can now construct the generalised geometry on the differential algebra $(\mathcal{A}_n, V)$. In the abstract-index notation, the metric $\mathcal{G}$ is represented by the doubly indexed array $g_{AB}$ where

$$g_{AB} := \mathcal{G}(\partial_A, \partial_B).$$

One can regard $g_{AB}$ as a square, symmetric and nonsingular matrix of order $m + n^2 - 1$. Its inverse matrix is denoted by $g^{AB}$. Exactly as in the standard differential geometry, the metric matrix and its inverse can be used to lower and raise indices of other multi-indexed entities.

We are now ready to define the *Levi-Civita connection* $\nabla : V \times V \to V$ by means of the Koszul formula. Namely, for any $u, v \in V$, $\nabla_u v$ is the unique derivation which satisfies

$$\begin{aligned}\mathcal{G}(\nabla_u v, w) := &\frac{1}{2}[u(\mathcal{G}(v,w)) + v(\mathcal{G}(u,w)) - w(\mathcal{G}(u,v)) \\ &+ .\mathcal{G}(w,[u,v]) + \mathcal{G}(v,[w,u]) - \mathcal{G}(u,[v,w])].\end{aligned}$$

for all $w \in V$. The connection's components (called the Christoffel symbols of the second kind) are defined by the equality

$$\nabla_{\partial_C}\partial_B = \Gamma^A_{BC}\partial_A$$

and by the Koszul formula they can be expressed as

$$\Gamma^A_{BC} = \frac{1}{2}g^{AD}(\partial_C g_{DB} + \partial_B g_{DC} - \partial_D g_{BC} + \mathbf{c}_{CBD} + \mathbf{c}_{DCB} - \mathbf{c}_{BDC}).$$

Due to the noncommutativity, we have

$$\Gamma^A_{BC} - \Gamma^A_{CB} = \mathbf{c}^A_{CB}.$$

Therefore, unlike the components of the standard Levi-Civita connection, $\Gamma^A_{BC}$ might not be symmetric with respect to the latter two indices.

In spite of this asymmetry, the Levi-Civita connection enjoys many of the properties of its standard counterpart, in particular, it is *torsion-free*, that is

$$\forall u, v \in V \quad \nabla_u v - \nabla_v u - [u, v] = 0$$

and *compatible with the metric*, that is

$$\forall u, v, w \in V \quad w(\mathcal{G}(u,v)) = \mathcal{G}(\nabla_w u, v) + \mathcal{G}(u, \nabla_w v)$$

Moreover, just like in the standard case, the Levi-Civita connection is the unique map $V \times V \to V$ satisfying the above two conditions.

The *Riemann curvature tensor* $R : V \times V \times V \to V$, $(u, v, w) \mapsto R(u, v)w$ is defined by means of $\nabla$ via

$$R(u,v)w := \nabla_u \nabla_v w - \nabla_v \nabla_u w - \nabla_{[u,v]} w.$$

Its components $R^C_{\ DAB}$ are defined through

$$R(\partial_A, \partial_B)\partial_D = R^C_{\ DAB}\partial_C,$$

and are given by the formula

$$R^C_{\ DAB} = \partial_A \Gamma^C_{\ DB} - \partial_B \Gamma^C_{\ DA} + \Gamma^K_{\ DB}\Gamma^C_{\ KA} - \Gamma^K_{\ DA}\Gamma^C_{\ KB} - c^K_{AB}\Gamma^C_{\ DK}.$$

This is nothing but the standard expression for the components of the Riemann tensor with an additional term that can be regarded as due to noncommutativity. It can be easily checked that $R$ enjoys the usual Riemann tensor symmetries

$$R_{CDAB} = -R_{DCAB} = -R_{CDBA} = R_{ABCD},$$
$$R^C_{\ DAB} + R^C_{\ BDA} + R^C_{\ ABD} = 0.$$

Finally, the Ricci tensor, $\mathbf{ric} : V \times V \to C^\infty(M)$ and the curvature scalar $r \in C^\infty(M)$ can be introduced as suitable contractions of the Riemann tensor. Concretely, the components of the Ricci tensor read

$$\mathbf{ric}_{AB} := R^C_{\ ACB},$$

whereas the curvature scalar is

$$r := g^{AB}\mathbf{ric}_{AB} = g^{AB} R^C_{\ ACB}.$$

Notice that, just as in the standard case, $\mathbf{ric}_{AB} = \mathbf{ric}_{BA}$.

With the generalised curvature tensors defined, we are ready to formulate and study the generalised vacuum Eistein equations.

## Generalised Einstein Equations

The standard derivation of the Einstein equations is conducted by means of the action principle, starting from a suitably chosen action functional. In standard GR, the so-called Einstein-Hilbert action is the integral of the curvature scalar over the entire space-time manifold. Therefore, the natural candidate for the generalised Einstein-Hilbert action is

$$S_{EH} := \int r\sqrt{|g|}\,d^m x, \tag{1}$$

where $m = \dim M$ and $g$ denotes the determinant of the metric matrix $g_{AB}$. We want to study the vacuum equations and therefore postulate no additional matter term.

The action principle amounts here to varying $S_{EH}$ with respect to $\delta g^{AB}$ and assuming that the variation vanishes. The calculation leads to the generalised Einstein equations of the form

$$\mathbf{ric}_{AB} = 0. \tag{2}$$

Although these equations look similar to the standard vacuum Einstein equations, they actually have a far richer content when projected onto space-time $M$. This is due to the extra terms coming from additional components of the metric (the "inner degrees of freedom"). These extra terms can be interpreted as an "$m$-dimensional matter-energy" induced by the generalised (vacuum) Einstein equations (similarly as it is the case in the Kaluza-Klein-type theories (Appelquist et al. 1987; Clifton et al. 2012; Lee 1984; Overduin and Wesson 1997), although without introducing extra geometrical dimensions). Let us consider an example of the following block diagonal metric

$$g_{AB} = \begin{bmatrix} g_{\mu\nu} & 0 \\ 0 & g_{\{A\}\{B\}} \end{bmatrix}.$$

Let us remember that $g_{\mu\nu}$ is an $m$-by-$m$ matrix and $g_{\{A\}\{B\}}$ is a $(n^2 - 1)$-by-$(n^2 - 1)$ matrix, where $n = |G|$. The inverse metric matrix $g^{AB}$ is also block diagonal.

Such block diagonal metric has an important feature: its corresponding Levi-Civita connection $\nabla$ *extends the standard Levi-Civita connection* $\tilde{\nabla}$ : $\operatorname{Der} C^\infty(M) \times \operatorname{Der} C^\infty(M) \to \operatorname{Der} C^\infty(M)$ in the sense that

$$\forall \bar{X}, \bar{Y} \in \operatorname{Hor} A_n \quad \nabla_{\bar{X}} \bar{Y} = \overline{\tilde{\nabla}_X Y}.$$

In other words, the Levi-Civita connection acts on horizontal derivations in the same way as does its classical counterpart.

For the block diagonal metric, the generalised Einstein equations can be written more explicitly than (2) as

$$\widetilde{\mathbf{ric}}_{\mu\nu} = \frac{1}{4}\Big(\tilde{\nabla}_\mu \tilde{\nabla}_\nu \ln|\breve{g}| + g^{\{A\}\{B\}} \tilde{\nabla}_\mu \tilde{\nabla}_\nu g_{\{A\}\{B\}}\Big), \tag{3}$$

$$\mathbf{c}_{\{B\}}^{\{C\}\{D\}} \partial_\mu g_{\{C\}\{D\}} = 0, \tag{4}$$

$$\begin{aligned} &\tilde{\Delta} g_{\{A\}\{B\}} - g_{\{A\}\{C\}} g_{\{B\}\{D\}} \tilde{\Delta} g^{\{C\}\{D\}} + \partial^\mu \ln|\breve{g}| \partial_\mu g_{\{A\}\{B\}} \\ &= -2c_{\{A\}}^{\{C\}\{D\}} \big(\mathbf{c}_{\{B\}\{C\}\{D\}} + \mathbf{c}_{\{B\}\{D\}\{C\}}\big) + \mathbf{c}_{\{A\}}^{\{C\}\{D\}} \mathbf{c}_{\{C\}\{D\}\{B\}}. \end{aligned} \tag{5}$$

where

- $\widetilde{\mathbf{ric}}_{\mu\nu}$ denotes the standard Ricci tensor.
- $\tilde{\nabla}_\mu$ is the covariant derivative resulting from the standard Levi-Civita connection $\tilde{\nabla}$. Note that $\tilde{\nabla}_\mu$ by definition "sees" only the lowercase Greek indices.
- $\tilde{\Delta} := g^{\mu\nu} \tilde{\nabla}_\mu \tilde{\nabla}_\nu$ is the standard Laplace-Beltrami operator.
- $\breve{g} := \det g_{\{A\}\{B\}}$.

Equation (3) is a projection of the generalised Einstein equations (2) onto the $m$-dimensional space-time $M$. Since it implies that

$$\tilde{r} = \frac{1}{4}\left(\tilde{\Delta}\ln|\breve{g}| + g^{\{A\}\{B\}} \tilde{\Delta} g_{\{A\}\{B\}}\right), \tag{6}$$

where $\tilde{r}$ denotes the standard curvature scalar, therefore we can equivalently write (3) in the form of the standard Einstein equations with a certain nonzero energy-momentum tensor

$$\begin{aligned} \tilde{G}_{\mu\nu} = \frac{1}{4}\Big[&\Big(\tilde{\nabla}_\mu \tilde{\nabla}_\nu - \frac{1}{2} g_{\mu\nu} \tilde{\Delta}\Big) \ln|\breve{g}| \\ &+ g^{\{A\}\{B\}} \Big(\tilde{\nabla}_\mu \tilde{\nabla}_\nu - \frac{1}{2} g_{\mu\nu} \tilde{\Delta}\Big) g_{\{A\}\{B\}}\Big], \end{aligned} \tag{7}$$

where $\tilde{G}_{\mu\nu} := \widetilde{\mathbf{ric}}_{\mu\nu} - \frac{1}{2} g_{\mu\nu} \tilde{r}$ is the standard Einstein tensor.

The fact that there appears a nonzero energy-momentum tensor can be regarded as a realisation of the "matter out of geometry" mechanism (Heller et al. 2007) or, in this case more precisely, of the "scalar fields out of noncommutative geometry" mechanism. One can thus regard Eqs. (4, 5) as the equations of state of those "emergent" scalar fields.

Additionally, we can also rewrite the generalised Einstein-Hilbert action (1) more explicitly. Namely

$$S_{EH} = \int \left[ \sqrt{|\breve{g}|}\tilde{r} - \frac{1}{4}\sqrt{|\breve{g}|}\left(\widetilde{\Delta}\ \ln|\breve{g}| + g^{\{A\}\{B\}}\widetilde{\Delta}g_{\{A\}\{B\}} + \mathbf{C}\right)\right]\sqrt{-\tilde{g}}d^m x,$$

where

- $\tilde{g} := \det g_{\mu\nu}$.
- $\mathbf{C} := \mathbf{c}^{\{B\}\{C\}\{D\}}\left(2\mathbf{c}_{\{B\}\{D\}\{C\}} + \mathbf{c}_{\{B\}\{C\}\{D\}}\right)$.

Notice that noncommutativity enters into the action both through the additional components of the metric (i.e. the "inner derivation block" $g_{\{A\}\{B\}}$) as well as through the nonzero structure constants.

One can regard the theory, obtained in this way, as an example of a scalar-tensor theory,[3] which involves no less than $n^2(n^2-1)/2$ independent scalar fields. These fields, when arranged into a symmetrical matrix $g_{\{A\}\{B\}}$, are such that $\breve{g} := \det g_{\{A\}\{B\}}$ is a nowhere vanishing field. Notice, moreover, that $\sqrt{|\breve{g}|}$ seems to play a special role in this theory, as it can be shown to satisfy the following Klein-Gordon-like equation

$$\left(\widetilde{\Delta} + \frac{1}{4}\mathbf{C}\right)\sqrt{|\breve{g}|} = 0 \tag{8}$$

with $\frac{1}{4}\mathbf{C}$ playing the role of the mass term. This might constitute another illustration of how noncommutative geometry gives rise to (massive) scalar fields in this model.

We now move to presenting two explicit solutions of Einstein Eqs. (3–5) in the simplest case when the structure group $G$ has only two elements. We restrict ourselves to "Friedman-like" solutions, by which we mean metrics, whose "horizontal derivation block" $g_{\mu\nu}$ is of the Friedman-Lemaitre-Robertson-Walker form and whose remaining components are time-dependent only[4]

$$g_{\mu\nu} = \begin{bmatrix} -1 & 0 & 0 & 0 \\ 0 & \frac{a^2(t)}{1-kr^2} & 0 & 0 \\ 0 & 0 & a^2(t)r^2 & 0 \\ 0 & 0 & 0 & a^2(t)r^2\sin^2\theta \end{bmatrix},$$

$$g_{\{A\}\{B\}} = g_{\{A\}\{B\}}(t).$$

Recall that $a(t)$ is called the *scale factor* and $k \in \{-1, 0, 1\}$ is the *curvature constant.*

To simplify the calculations, we assume furthermore that the "inner derivation block" $g_{\{A\}\{B\}}$ has the following form

---

[3]See e.g. (Clifton et al. 2012) and references therein.

[4]As usual, we set $c = G = 1$.

$$g_{\{A\}\{B\}}(t) = \begin{bmatrix} \xi f^2(t) & 0 & 0 \\ 0 & 0 & \eta f^2(t) \\ 0 & \eta f^2(t) & 0 \end{bmatrix},$$

where $\xi, \eta \in \mathbb{R}\backslash\{0\}$ and $f$ is a time-dependent nowhere vanishing function.

One of the Einstein equations, (4), is satisfied automatically. The remaining two equations (3, 5) yield the following overdetermined nonlinear system of ordinary differential equations

$$\begin{cases} \frac{\ddot{a}}{a} + \frac{\ddot{f}}{f} = 0, \\ a\ddot{a}f + 2\dot{a}^2 f + 3a\dot{a}\dot{f} = -2kf, \\ f\ddot{f}a + 2\dot{f}^2 a + 3f\dot{f}\dot{a} = \frac{1}{\eta}a. \end{cases} \tag{9}$$

together with an additional algebraical condition that $\xi = 2\eta$.

Notice that, by the last of the above equations the function $f$ cannot be constant. Moreover, the Hubble parameter $H$ can be expressed entirely in terms of $f$ and $\eta$ via

$$H := \frac{\dot{a}}{a} = \frac{\eta^{-1} - f\ddot{f} - 2\dot{f}^2}{3f\dot{f}}.$$

Two explicit solutions of system (9),[5] one for $k = 0$ and another for $k = -1$, read:

$$\text{for } k = 0 \quad a(t) = a_0, \quad f(t) = \tfrac{1}{\sqrt{2\eta}}(t - t_0), \tag{10}$$

$$\text{for } k = -1 \quad a(t) = \sqrt{\tfrac{2}{5}}(t - t_0), \quad f(t) = \tfrac{1}{\sqrt{5\eta}}(t - t_0), \tag{11}$$

where $\eta, a_0, t_0$ are constants. In fact, without any loss of generality one can take $\eta = 1$ and $t_0 = 0$. Note that $g_{AB}$ becomes degenerate at $t = 0$.

First solution (10) describes the flat Minkowski space-time, although it is *not* a static solution since $\dot{f} \neq 0$ (and the above mentioned degeneracy occurs at $t = 0$).

Second solution (11) describes a hyperbolic, linearly expanding universe with the initial singularity at $t = 0$, resembling (but different from) the metric studied first by Milne (1933), in which $k = -1$ but $a(t) = t$. Milne's unconventional cosmology has recently gained a renewed interest in the work of Benoit-Lévy and Chardin (2012) in the form of the so-called Dirac-Milne universe, which is argued to be a viable alternative to the ΛCMD model. What is noteworthy, both the Milne model and the model governed by (11) are free from the cosmic age problem and from the horizon problem, where the latter is solved without introducing inflation (Benoit-Lévy and Chardin 2012).

[5]There might be more, however finding them (or proving that other solutions do not exist) seems like a daunting task!

However, there is a significant difference between Milne's metric and solution (11). Namely, Milne's universe is devoid of energy-matter; it is a vacuum solution. On the other hand, solution (11) is associated to a nonzero energy-matter tensor. Since the solution is Friedman-like, the energy-matter tensor describes some sort of a perfect fluid. Let us therefore see, *what kind of perfect fluid is in this case induced by the (noncommutative) geometry.*

The standard Einstein tensor assumes here the following (mixed) form

$$\widetilde{G}^{\mu}_{\ \nu} = \frac{3}{2t^2}\mathrm{diag}(3, 1, 1, 1)$$

Therefore, by Einstein equations, the induced perfect fluid energy density $\rho$ and pressure $p$ are

$$\rho(t) = -\frac{36\pi}{t^2}, \quad p(t) = \frac{12\pi}{t^2}.$$

Note that $\rho$ is negative and therefore in this case no classical fluid matches the induced one. In fact, such fluid would violate various energy conditions of general relativity.

Nevertheless, Friedman cosmology involving negative $\rho$ has been studied in (Nemiroff et al. 2015), where it is argued that some realisations and extensions of quantum field theories do allow for such exotic energy forms, albeit only locally. In the article cited, the perfect fluid with $\rho < 0$ and $w = p/\rho = -1/3$ is referred to as *negative cosmic strings*. The interplay between such *negative* forms of energy and the classical *positive* ones leads to interesting cosmological scenarios, also studied in Nemiroff et al. (2015).

## Concluding Remarks

How do the models constructed in the present work inscribe into philosophies (alluded to in the Introduction) concerning the relationship between matter and space-time geometry? It is straightforward that they nicely fall into the "matter out of geometry" heading. Original Wheeler's program could not succeed since the standard space-time geometry had not enough degrees of freedom to accommodate his postulates. As our work has demonstrated, noncommutative generalisation of geometry creates such possibilities. The remarkable fact is that not only the energy-momentum tensor is recovered from purely geometric degrees of freedom (without postulating additional space dimensions), but also suitable equations of state can be obtained in this way. The fact that these astonishing properties are produced in the framework of very simplified models allows us to expect even more interesting effects when more realistic models are constructed.

The main feature of noncommutative generalisation of geometry consists of a strong shift in the interplay between local and global geometric properties. In strongly noncommutative spaces, local properties are entirely engulfed by their global structure. Such spaces are nonlocal entities, in the sense that local notions in them are, in general, devoid of meaning and can only be recovered as some "limiting cases". Our algebra $\mathcal{A}$ of smooth compactly supported functions on the groupoid $\Gamma$ of Lorentz symmetries with noncommutative multiplication determines a noncommutative space with milder properties. Local properties are virtually present in it as encoded in the fact that the center $\mathcal{Z}(\mathcal{A})$ of the algebra $\mathcal{A}$ is nontrivial. Owing to this fact, the usual geometry of space-time, with all its local properties, can naturally be recovered.

And here we have a contact with Mach's principle. As we remember, according to its strong version all local properties of space-time should be entirely determined by space-time global structure. This is exactly what happens in our models. On the level of the algebra $\mathcal{A}$, the global structure is dominating the scene, and it is this structure that determines, through the center of $\mathcal{A}$, all local properties of space-time $M$. We could say that the noncommutative space as determined by the algebra $\mathcal{A}$ is fully Machian, and all anti-Machian properties of space-time $M$ emerge when "noncommutative symmetries" are broken to the usual space-time symmetries.

In this way, the "ugly dichotomy" of space-time and matter is removed, and Einstein's starving for a monistic vision of the universe could be satisfied, albeit on the level of simplified models.

## References

Appelquist, T., Chodos, A., & Freund, P. G. O. (Eds.). (1987). *Modern Kaluza-Klein theories*. Menlo Park: Addison-Wesley.

Barbour, J., & Pfister, H. (Eds.). (1995). *Mach's Principle. From Newton's Bucket to Quantum Gravity*. Boston, Basel, Berlin: Birkhäuser.

Benoit-Lévy, A., & Chardin, G. (2012). Introducing the Dirac-Milne universe. *Astronomy & Astrophysics, 537,* A78.

Chamseddine, A. H., Connes, A., & Marcolli, M. (2007). Gravity and the standard model with neutrino mixing. *Theoretical and Mathematical Physics, 11,* 991–1089.

Clifton, T., Ferreira, P. G., Padilla, A., & Skordis, C. (2012). Modified Gravity and Cosmology. *Physics Reports, 513*(1), 1–189.

Connes, A. (1994). *Noncommutative geometry*. New York: Academic Press.

Connes, A. (2006). *Noncommutative geometry and physics*, http://www.alainconnes.org/docs/einsymp.pdf.

Heller, M., Miller, T., Pysiak, L., & Sasin, W. (2015). Geometry and general relativity in the groupoid model with a finite structure group. *Canadian Journal of Physics, 93*(1), 43–54.

Heller, M., Pysiak, L., & Sasin, W. (2005). Noncommutative dynamics of random operators. *International Journal of Journal of Theoretical Physics, 44*(6), 619–628.

Heller, M., Pysiak, L., & Sasin, W. (2007). Conceptual unification of gravity and quanta. *International Journal of Theoretical Physics, 46*(10), 2494–2512.

Heller, M., & Sasin, W. (1999). Noncommutative unification of general relativity and quantum mechanics. *International Journal of Theoretical Physics, 38,* 1619–1642.

Heller, M., Sasin, W., & Lambert, D. (1997). Groupoid approach to noncommutative quantization of gravity. *Journal of Mathematical Physics, 38,* 5840–5853.

Jureit, J.-H., Krajewski, T., Schucker, T., & Stephan, C. A. (2007). On the noncommutative standard model. *Acta Physica Polonica B, 38,* 3181–3202.

Lee, H. C. (Ed.). (1984). An introduction to Kaluza-Klein theories. In *Proceedings of Chalk River Workshop on Kaluza-Klein Theories*. Singapore: World Scientific.

Milne, A. E. (1933). World-structure and the expansion of the universe. *Zeitschrift fur Astrophysik, 6,* 1–95.

Nemiroff, R. J., Joshi, R., & Patla, B. R. (2015). An exposition on Friedmann cosmology with negative energy densities. *Journal of Cosmology and Astroparticle Physics, 06,* 006.

Overduin, J. M., & Wesson, P. S. (1997). Kaluza-Klein Gravity. *Physics Reports, 283*(5–6), 303–378.

Penrose, R., & Rindler, W. (1984). *Spinors and space-time* (Vol. 1). Cambridge: Cambridge University Press.

Wheeler, J. A. (1964). Geometrodynamics and the issue of the final state. In C. DeWitt & B. DeWitt (Eds.), *Relativity, groups and topology* (pp. 315–520). Gordon and Breach.

Wheeler, J. A. (1968). Superspace and the nature of quantum geometrodynamics. In C. DeWitt & J. A. Wheeler (Eds.), *Battelle Rencontres* (pp. 242–307). Benjamin.

Wheeler, J. A. (1973). From relativity to mutability. In J. Mehra (Ed.), *The Physicist's Conception of Nature* (pp. 202–247). D. Reidel.

# Testing New Physics with Polarized Light: Cosmological Birefringence

Matteo Galaverni

**Abstract** Cosmological rotation of linear polarization (cosmic birefringence or cosmic polarization rotation) provides an excellent probe to study new physics. We derive stringent limits on few selected extensions of the Standard Model looking at the linear polarization rotation angle for several datasets (Cosmic Microwave Background, Radio Galaxies, Radio Sources, Crab Nebula, Gamma-ray Bursts) corresponding at different energies for the photons and different distances of the sources.

## Introduction

Nowadays many compelling evidences for new physics beyond the Standard Model are provided by astrophysical and cosmological observations (e.g. dark matter, dark energy, …). Several extensions of the Standard Model are currently under study from breaking of fundamental symmetries (e.g. Lorentz symmetry) to modifications of General Relativity. We try to constrain these theories looking their effects on propagation of light. We focus in particular on cosmological rotation of linear polarization (cosmological birefringence or cosmic polarization rotation) for different kinds of sources (Cosmic Microwave Background, Radio Galaxies, Radio Sources, Crab Nebula and Gamma-ray Bursts).

In this work, we use natural units, $\hbar = c = 1$, and assume a cosmological model with *Planck* 2013 estimates of cosmological parameters (Ade et al. 2014) $H_0 = h \cdot 100$ km/s/Mpc = 67.2 km/s/Mpc, $\Omega_M = 0.32$ and $\Omega_\Lambda = 0.68$. The amount of rotation of the polarization plane is denoted by $\alpha$.

M. Galaverni (✉)
Studio Teologico Interdiocesano, V.le Timavo 93, 42121 Reggio Emilia, Italy
e-mail: matteo.galaverni@gmail.com



G. Gionti, S.J. and J.-B. Kikwaya Eluo, S.J. (eds.), *The Vatican Observatory, Castel Gandolfo: 80th Anniversary Celebration*, Astrophysics and Space Science Proceedings 51, https://doi.org/10.1007/978-3-319-67205-2_11

## Polarization Conventions

Polarization of light is usually described in terms of Stokes parameters: Q and U for linear polarization, V for circular polarization.

Before combining the different constraints for the linear polarization rotation angle $\alpha$ we remember the existence of two polarization conventions. Indeed the definition of the Stokes parameters Q and U depends on the coordinate system used. Two are the possible definitions:

- IAU/IEEE[1]: in each point of the sky sphere is defined a right-handed reference system with $x$ axis points toward North, $y$ points toward East, and $z$-axis points inward (toward the observer). Therefore looking toward the source the linear polarization angle increases anti-clockwise (see Fig. 1) (Hamaker and Bregman 1996).
- HEALPix[2]: in each point of the sky sphere is defined a right-handed reference system with $x$ axis points toward South, $y$ points toward East, and $z$-axis points outward (toward the source). Therefore looking toward the source the linear polarization angle increases clockwise (see Fig. 2) (Gorski et al. 1999).

Because different communities are involved in this kind of measurements attention is needed (Hamaker and Leahy 2004; Ade 2015; Galaverni and Finelli 2007). Since this review is mainly based on Galaverni et al. (2015) we adopt the HEALPix conventions widely used for Cosmic Microwave Background observations.

## Dataset

We combine here several datasets by taking into account the different energies of the photons and distances of the sources.

### *Cosmic Microwave Background*

Polarization of the Cosmic Microwave Background (CMB) was predicted soon after its discovery (Rees 1968): Thomson scattering of unpolarized photons at last scattering surface generates linear polarization if the incident intensity varies with direction (anisotropies).

Polarization measurements are quite challenging: since the level of polarization is expected to be 1–10% of the amplitude of the temperature anisotropies depending

[1]International Astronomical Union/Institute of Electrical and Electronics Engineers.

[2]Hierarchical, Equal Area, and iso-Latitude Pixelisation of the sphere (Gorski et al. 1999).

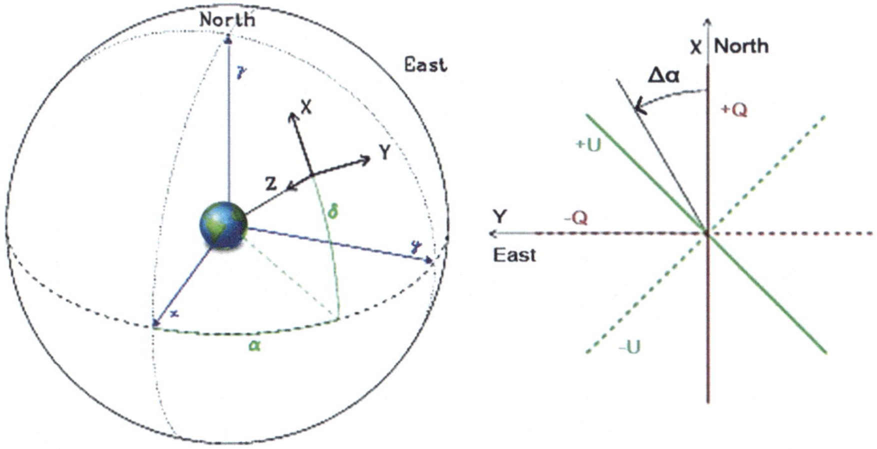


**Fig. 1** (Left) A schematic illustration of IAU/IEEE coordinate conventions. (Right) The linear polarization angle increases anti-clockwise. See also Gorski et al. (1999), Hamaker and Leahy (2004)

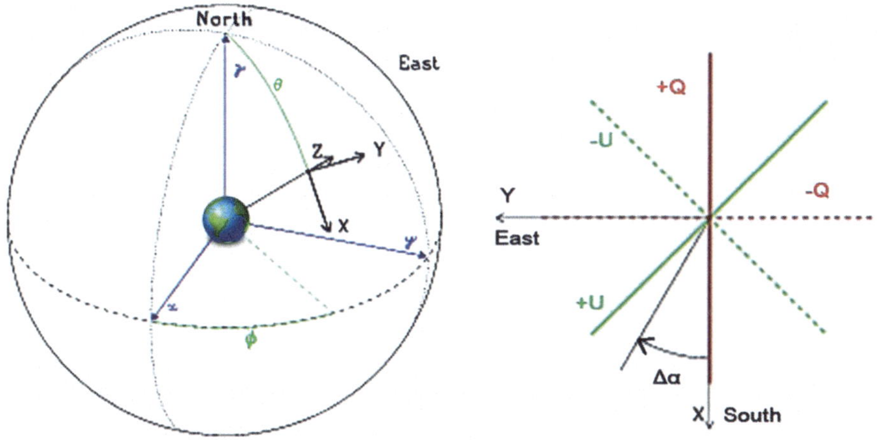


**Fig. 2** (Left) A schematic illustration of HEALPix coordinate conventions. (Right) The linear polarization angle increases clockwise

on the angular scale. The first detection was made by the Degree Angular Scale Interferometer (DASI) in 2002 (Kovac et al. 2002), in 2006 WMAP (Wilkinson Microwave Anisotropy Probe) (Page et al. 2007) released the first full sky maps. Several experiments (ground and satellites) confirmed polarization detection and currently the best full sky maps are provided by the Planck satellite (Adam et al. 2015).

CMB constraints for cosmological birefringence can be combined in several ways (Gubitosi and Paci 2013; Kaufman et al. 2015). For our purposes, it is enough to use the constraint from WMAP9 (Hinshaw et al. 2013). By adding in quadrature statistical and systematic errors, we obtain a constraint on cosmological birefringence $\alpha_{\text{CMB}} = -0.36 \pm 1.9$ deg for photons observed at an energy of $2.2 \times 10^{-4}$ eV.

## *UV Emission from Radio Galaxies*

Perpendicularity between the optical/UV axis and the linear optical/UV polarization of distant Radio Galaxies can also provide good constraints on $\alpha$. First limits obtained in early Nineties (Cimatti et al. 1994) have been recently updated (di Serego Alighieri et al. 2010). Averaging on different sources (average distance $z = 2.62$) we obtain the following constrain $\alpha = 0.7 \pm 2.1$ deg for photons with average energy of few eV (2.5 eV).

## *Radio Sources*

Another constraint on cosmological birefringence can be derived from the relation between polarization and total intensity structures of radio galaxies and quasars (Leahy 1997): $\alpha = 1.6 \pm 1.8$ deg ($E \simeq 3.4 \times 10^{-5}$ eV and $z = 0.47$).

## *Crab Nebula*

At much higher energies than the ones considered until now, constraints based on the observation of the Crab Nebula have been set ($E \sim 10^5$ eV); we refer in particular to Maccione et al. (2008). The distance of the Crab Nebula is 1.9 Kpc, corresponding to $z = 4.5 \times 10^{-7}$. Limits on the birefringence angle are of the order of few degrees: $\alpha = 1 \pm 11$ (in the CMB convention).

## *Summarising: Our Dataset*

We considered birefringence constraints over a very large energy (from few GHz to $10^9$ GHz, from few $\mu$ eV to hundreds of keV) and distance range (from z of few $10^{-7}$ to 1100). In order to clarify the specific contribution of each dataset we

**Table 1** Current constraints on the cosmological birefringence angle $\alpha$ coming from a variety of astrophysical and cosmological observations; for each dataset we report the typical redshift and the effective energy

| Dataset | $z$ | $E$ (eV) | $\alpha \pm \Delta\alpha$ (deg) | Reference |
|---|---|---|---|---|
| CMB | 1090 | $2.2 \times 10^{-4}$ | $-0.36 \pm 1.9$ | Hinshaw et al. (2013) |
| UV RG | 2.62 | 2.5 | $0.7 \pm 2.1$ | di Serego Alighieri et al. (2010) |
| Radio sources | 0.47 | $3.4 \times 10^{-5}$ | $1.6 \pm 1.8$ | Leahy (1997) |
| Crab Nebula | $4.5 \times 10^{-7}$ | $2.3 \times 10^{5}$ | $1 \pm 11$ | Maccione et al. (2008) |

decided to consider only a representative limit for each single data set (see Table 1 for a summary).

## Analysis and Results

In this section we discuss several models predicting birefringence, each one characterized by a different energy dependence for $\alpha$.

### *Energy-Independent Rotation*

Here, we will concentrate on an energy-independent birefringence effect linearly dependent on the propagation distance (Chern-Simons term (Carroll et al. 1990), coupling between the electromagnetic field and a scalar (quintessential) field (Carroll 1998), ...). Taking into account the universe expansion the expected amount of rotation $\alpha$ is:

$$\alpha(z_*) = -\frac{1}{2} p_0 \int_0^{z_*} \frac{1}{(1+z)H(z)} dz. \tag{1}$$

where $z_*$ is the source redshift and $p_0$ is the time-component of a fixed time-like vector which is coupled to the electromagnetic field.[3] Combining all data we obtain:

$$p_0 = (-0.93 \pm 2.9) \times 10^{-35} h \text{ eV}, \tag{2}$$

[3]In principle, as done in Carroll et al. (1990), one could consider a general vector $p_\alpha$, but this would produce non-isotropic effects. Here we work under the assumption that birefringence is isotropic, so that only the time part of the vector could be present in the model.

at 68% C.L. Note that the dominant contribution to the result comes from the CMB and UV radio galaxies, because of their significantly higher distance.

## *Energy-Dependent Rotation*

In this case our dataset of Table 1 can be further extended. We rely not only on a direct measurement of the cosmological birefringence, but limits are derived also from linear polarization measurements at different energies.

At high energies limits can be derived from polarization measurements of Gamma-ray bursts. We refer in particular as an example of the GRB capabilities to GRB061122 (Gotz et al. 2013), as this provides one of the latest results. The polarization direction is measured in two different energy bands, 250–350 and 350–800 keV, obtaining, respectively, $\phi_1 = 145 \pm 15$ and $\phi_2 = 160 \pm 20$ at 68% C.L. The distance of the GRB source is given as $z = 0.54$. As done in Gotz et al. (2013), we will use the conservative constraint on the rotation angle $\alpha = 0 \pm 50$ degrees (68% C.L.).[4]

At lower energies limits on the energy dependence of the linear polarization angle can be obtained looking at Mars polarized emission measured at different wavelengths (Perley and Butler 2013).

### Linear Energy Dependence

This dependence can be due to the 'Weyl' interaction described in Shore (2005) and Kahniashvili and Durrer (2008). The polarization rotation angle depends linearly on the distance travelled by photons, $\Delta\ell$, and on the dimensionless scalar $\Psi_0$:

$$\alpha(E_0, z_*) = 8\pi E_0 \Psi_0 \int_0^{z_*} H(z)^{-1} dz, \tag{3}$$

where $E_0$ is the photon energy today. Combining the first four datasets of Table 1, the best-fit value for $\Psi_0$ is:

$$\Psi_0 = (3.0 \pm 9.1) \times 10^{-37} h, \tag{4}$$

[4]Following (Gubitosi and Paci 2013) we introduce an effective energy depending on the functional dependence of $\alpha$ on energy and on the bandwidth of the different channels: $E = 530$ keV for the linear dependence and $E = 550$ keV in the quadratic case.

at 68% C.L.; the dominant contribution comes from UV radio galaxies. This is an interesting example of a case in which the dominant contribution does not come from the highest energetic source (the Crab Nebula), that is what one might naively expect. In fact, the distance dependence plays an important role and can compensate for the lower energy of other more distant sources (see Fig. 3). If we include also the constraint from GRB, this dominates and we obtain, at 68% C.L.: $\Psi_0 = (0.0 \pm 3.0) \times 10^{-40}h$. In both cases the constraint improves by several orders of magnitude the estimate based on CMB data only, $|\Psi_0| < 5.8 \times 10^{-33}h$ (Gubitosi and Paci 2013).

## Quadratic Energy Dependence

Quantum Gravity Planck-scale effects (Myers and Pospelov 2003; Gubitosi et al. 2009) can produce this kind of cosmological birefringence. If, following (Myers and Pospelov 2003; Gubitosi et al. 2009), we write the coupling constant between the EM field and the vector through a dimensionless parameter $\xi$ and the Planck mass scale $M_P$, then:

$$\alpha(E_0, z_*) = \frac{\xi}{M_P} E_0^2 \int_0^{z_*} (1+z)H(z)^{-1} dz. \quad (5)$$

Using this formula for our analysis, the best-fit estimate for $\xi$ is:

$$\xi = (1.2 \pm 14.1) \times 10^{-11}, \quad (6)$$

at 68% C.L. including all data points except GRB. Differently from what happened in the linear energy dependence, now it is actually the highest energy source (Crab Nebula) that gives the most important contribution, weighing the energy of the source more that its distance (see Fig. 3). If we include also the constraint from GRB we obtain at 68% C.L.: $\xi = (0.0 \pm 8.6) \times 10^{-17}$. As expected, the GRB provides the dominant contribution and our result is indeed compatible with the upper limit presented in Gotz et al. (2013). Again, in both cases the result improves the constraint, $\xi = (-0.22 \pm 0.22)$ at 68% C.L., based only on CMB dataset (Gubitosi and Paci 2013) by several orders of magnitude.

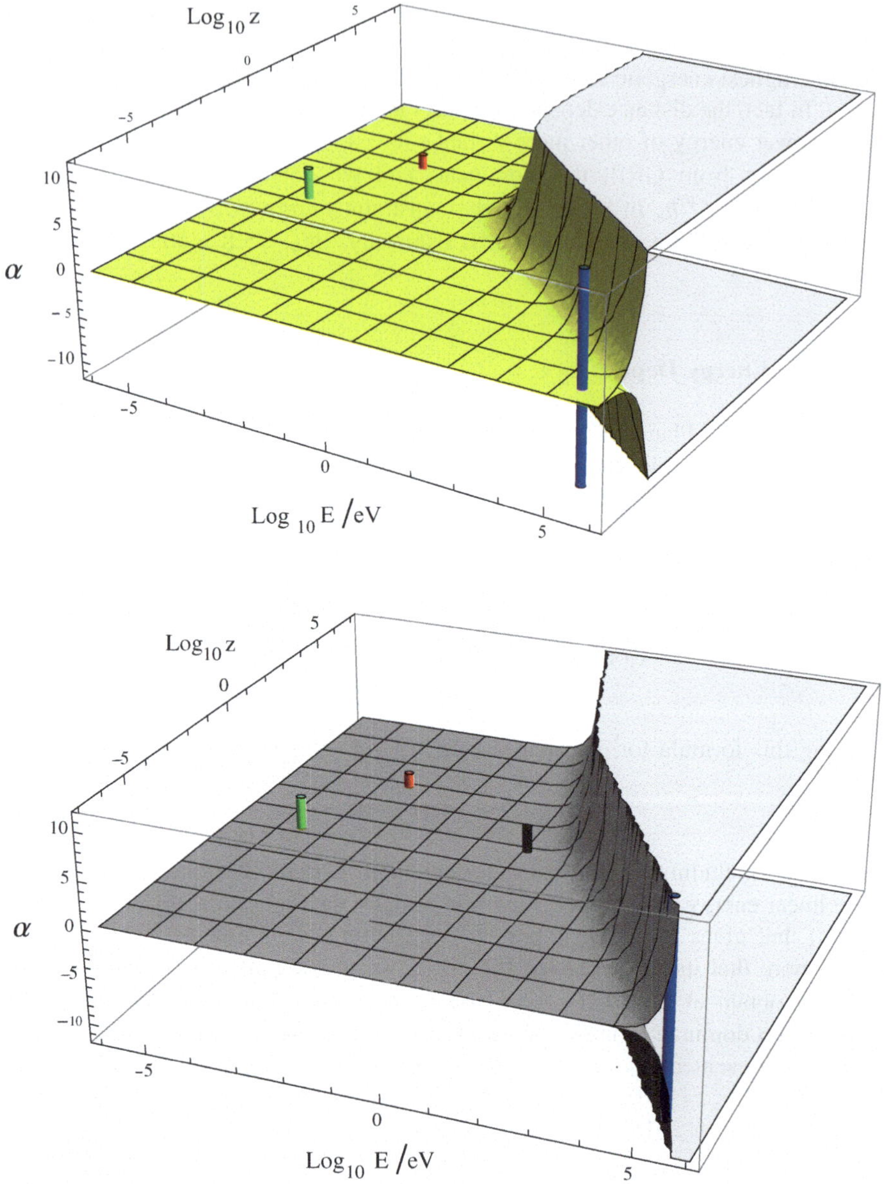


**Fig. 3** Constraints for the cosmological birefringence angle $\alpha$; black, green, red and blue points refer to UV radio galaxies, radio sources, CMB, Crab Nebula. (Upper panel): for linear energy dependence the dominant contribution comes from UV radio galaxies, we show in yellow the value of $\alpha(E, z)$ fixed $\Psi_0 = (3.0 \pm 9.1) \times 10^{-37} h$. (Lower panel): for quadratic energy dependence is Crab Nebula (highest energy source) that gives the most important contribution, we show in gray the value of $\alpha(E, z)$ fixed $\xi = (1.2 \pm 14.1) \times 10^{-11}$

## Conclusions

In the present work constraints on the rotation angle set by cosmological (CMB) and astrophysical (UV distant radio galaxies, radio sources, Crab Nebula, GRBs) observations were combined for the first time. Besides, updating current constraints on the models considered, this analysis provides also a useful guide for future polarization measurements aimed at investigating specific energy- and distance-dependent birefringence effects.

**Acknowledgements** The author is grateful to Giulia Gubitosi, Fabio Finelli, Jonathan P. Kaufman, Brian G. Keating and Sperello di Serego Alighieri for valuable comments on this work. Thanks to INAF-IASF Bologna and Vatican Observatory for hospitality during the development of this project.

## References

Adam, R., et al. (2016). [Planck Collaboration], Planck 2015 results. I. Overview of products and scientific results. *Astronomy and Astrophysics, 594*, A21. doi:10.1051/0004-6361/201527101 [arXiv:1502.01582 [astro-ph.CO]].

Ade, P.A.R., et al. (2014). [Planck Collaboration], Planck 2013 results. XVI. Cosmological parameters. *Astronomy and Astrophysics* [arXiv:1303.5076 [astro-ph.CO]].

Ade, P. A. R., et al. (2015). [Planck Collaboration], Planck intermediate results. XIX. An overview of the polarized thermal emission from Galactic dust. *Astronomy and Astrophysics, 576*, A104. doi:10.1051/0004-6361/201424082 [arXiv:1405.0871 [astro-ph.GA]].

Carroll, S. M. (1998). Quintessence and the rest of the world. *Physical Review Letters, 81*, 3067 [astro-ph/9806099].

Carroll, S. M., Field, G. B., & Jackiw, R. (1990). Limits on a Lorentz and parity violating modification of electrodynamics. *Physical Review D, 41*, 1231.

Cimatti, A., di Serego Alighieri, S., Field, G. B., & Fosbury, R. A. E. (1994). Stellar and scattered light in a radio galaxy at z = 2.63. *The Astrophysical Journal, 422*, 562.

di Serego Alighieri, S., Finelli, F., & Galaverni, M. (2010). Limits on cosmological birefringence from the UV polarization of distant radio galaxies. *The Astrophysical Journal, 715*, 33. doi:10.1088/0004-637X/715/1/33 [arXiv:1003.4823 [astro-ph.CO]].

Galaverni, M., & Finelli, F. (2007). Systematics of cosmic microwave background polarization. Internal Report IASF-BO 454.

Galaverni, M., Gubitosi, G., Paci, F., & Finelli, F. (2015). Cosmological birefringence constraints from CMB and astrophysical polarization data. *Journal of Cosmology and Astroparticle Physics, 1508*(08), 031. doi:10.1088/1475-7516/2015/08/031 [arXiv:1411.6287 [astro-ph.CO]].

Gorski, K. M., Wandelt, B. D., Hansen, F. K., Hivon, E., & Banday, A. J. (1999). The healpix primer [astro-ph/9905275].

Gotz, D., Covino, S., Fernandez-Soto, A., Laurent, P., & Bosnjak, Z. (2013). The polarized gamma-ray burst GRB 061122. *Monthly Notices of the Royal Astronomical Society, 431*, 3550. doi:10.1093/mnras/stt439 [arXiv:1303.4186 [astro-ph.HE]].

Gubitosi, G., & Paci, F. (2013). Constraints on cosmological birefringence energy dependence from CMB polarization data. *Journal of Cosmology and Astroparticle Physics, 1302*, 020 [arXiv:1211.3321 [astro-ph.CO]].

Gubitosi, G., Pagano, L., Amelino-Camelia, G., Melchiorri, A., & Cooray, A. (2009). A constraint on Planck-scale modifications to electrodynamics with CMB polarization data. *Journal of Cosmology and Astroparticle Physics, 0908*, 021 [arXiv:0904.3201 [astro-ph.CO]].

Hamaker, J. P., & Bregman, J. D. (1996). Understanding radio polarimetry. III. Interpreting the IAU/IEEE definitions of the Stokes parameters. *Astronomy and Astrophysics Supplement Series, 117*, 161.

Hamaker, J. P., & Leahy, J. P. (2004). A study of CMB differencing polarimetry with particular reference to Planck. ESA Report SCI-A/2003.312/JT.

Hinshaw, G., et al. (2013). [WMAP Collaboration], Nine year Wilkinson microwave anisotropy probe (WMAP) observations: Cosmological parameter results. *The Astrophysical Journal Supplement Series, 208*, 19. doi:10.1088/0067-0049/208/2/19 [arXiv:1212.5226 [astro-ph.CO]].

Kahniashvili, T., Durrer, R., & Maravin, Y. (2008). Testing Lorentz invariance violation with WMAP five year data. *Physical Review D, 78*, 123009 [arXiv:0807.2593 [astro-ph]].

Kaufman, J. P., Keating, B. G., & Johnson, B. R. (2015). Precision tests of parity violation over cosmological distances. *Monthly Notice of the Royal Astronomical Society, 455*(2), 1981. doi:10.1093/mnras/stv2348 [arXiv:1409.8242 [astro-ph.CO]].

Kovac, J., Leitch, E. M., Pryke, C., Carlstrom, J. E., Halverson, N. W., & Holzapfel, W. L. (2002). Detection of polarization in the cosmic microwave background using DASI. Nature, *420*, 772. doi:10.1038/nature01269 [astro-ph/0209478].

Leahy, J. P. (1997). Comment on the measurement of cosmological birefringence [astro-ph/9704285].

Maccione, L., Liberati, S., Celotti, A., Kirk, J. G., & Ubertini, P. (2008). Gamma-ray polarization constraints on Planck scale violations of special relativity. *Physical Review D, 78*, 103003 [arXiv:0809.0220 [astro-ph]].

Myers, R. C., & Pospelov, M. (2003). Ultraviolet modifications of dispersion relations in effective field theory. *Physical Review Letters*, *90*, 211601 [hep-ph/0301124].

Page, L., et al. (2007). [WMAP Collaboration], Three year Wilkinson microwave anisotropy probe (WMAP) observations: Polarization analysis. *The Astrophysical Journal Supplement Series, 170*, 335. doi:10.1086/513699 [astro-ph/0603450].

Perley, R. A., & Butler, B. J. (2013). Integrated polarization properties of 3C48, 3C138, 3C147, and 3C286. *The Astrophysical Journal Supplement Series, 206*, 16. doi:10.1088/0067-0049/206/2/16 [arXiv:1302.6662 [astro-ph.IM]].

Rees, M. J. (1968). Polarization and spectrum of the primeval radiation in an anisotropic universe. *The Astrophysical Journal*, *153*, L1.

Shore, G. M. (2005). Strong equivalence, Lorentz and CPT violation, anti-hydrogen spectroscopy and gamma-ray burst polarimetry. *Nuclear Physics B*, *717*, 86 [hep-th/0409125].

# Part V
# History of Astronomy, Education and Outreach

# Showing the World: How Vatican Astronomers Interact with the Popular Press

Br. Guy Consolmagno, S.J.

In 1891, when Pope Leo XIII established the Vatican Observatory, he did it for a specific reason: "…that everyone might see clearly that the Church and her Pastors are not opposed to true and solid science, whether human or divine, but that they embrace it, encourage it, and promote it with the fullest possible devotion." (from: *Motu Proprio*, "The Refounding and Restructuring of the Vatican Observatory", 14 March 1891). Obviously, our mission is to do true and solid science. But note that our mission is more than that; our mission is also to make certain that "everyone might see clearly" the science that we do, supported by the Church and her Pastors.

The task of making sure that everyone sees what we are doing goes by many names and takes many forms.

It includes public presentations in schools, astronomy clubs, universities, and other public forums like the Council on World Affairs or the British Science Festival (to name two groups that members of the Observatory have addressed in recent years). One particular way we can interact with schools and other groups is by hosting visitors and running tours of our facilities, something that will become much more regularized with the opening of the Visitors' Center now under construction in the old Schmidt and Carte du Ciel domes within the Papal Gardens here in Castel Gandolfo.

This outreach also involves programs that are specifically run by the Vatican Observatory Foundation (VOF) in the US, such as our annual Faith and Astronomy Workshops. This event invites 25 educators from Catholic parishes to spend a week in Tucson, getting to know local astronomers and touring astronomical sites such at the University of Arizona mirror lab and a behind-the-scenes visit to the telescopes on Kitt Peak.

The internet has provided a particularly useful tool in our outreach. The VOF has started an intense program of outreach to Catholic high schools in the US, including

Br. G. Consolmagno, S.J. (✉)
Specola Vaticana, Vatican City State, Holy See (Vatican City State)
e-mail: gjc@specola.va

 
G. Gionti, S.J. and J.-B. Kikwaya Eluo, S.J. (eds.), *The Vatican Observatory, Castel Gandolfo: 80th Anniversary Celebration*, Astrophysics and Space Science Proceedings 51, https://doi.org/10.1007/978-3-319-67205-2_12

visits via Skype (as well as in person) and getting students to use the Vatican Advanced Technology Telescope (VATT). We participate with the Jesuit Virtual Learning Academy, teaching an online introductory course in astronomy to high school students. And most notably, the VOF runs a blog site called The Catholic Astronomer, where thousands of readers every month check out daily postings from half a dozen different contributors, both members of staff and our lay and clerical collaborators, who write about our lives and passions as astronomers and Catholics (www.vofoundation.org/blog).

Last but hardly least are the significant number of articles and popular books written by members of the Observatory. This book is but one example of a large library of publications produced in many languages on many topics over the history of the Specola.

All of these activities are essential to our mission. I note that many of the examples I have given here are centered in the US at the moment and are conducted in English; that simply arises from the role that the US-based VOF plays in raising funding and other support for the VATT within the US. But members of the Observatory come from all over the world, and many of us carry on similar outreach and engagement activities in areas ranging from Eastern Europe to South America to Africa.

Such activities fit well with the sorts of things we astronomers are used to doing and are comfortable with doing: teaching, writing articles, sharing our enthusiasm for our work with groups of people who are already interested in what we do and want to know more. But there is one essential activity for engaging the public and showing the world the Church's support of science for which many of us feel particularly uncomfortable. That involves dealing with members of the press, including newspaper, radio, television, and internet interviews.

Unlike every other example, when dealing with an interviewer we cannot assume that the reporters, or their audience, are already on our side. The best we can expect is an initial indifference to our work. In fact, there may be times when either the interviewer or the audience might actually be hostile to who we are and what we have to say. Furthermore, unlike all the previous examples, when we are the people being interviewed we no longer are in control of the message or the way it will be presented. Thus, speaking to the press requires a set of skills that are foreign to our experience. For that reason, such interviews are prone to make the typical astronomer feel anxious, uncomfortable, and unwilling to participate. To help alleviate that unease, this chapter will concentrate specifically on a set of useful rules and principles for astronomers interacting with journalists.

The ideas here come from many sources. Many of them I learned, literally, at my father's knee: for many years he ran the press office of a large American corporation and he shared with me while I was a student (and at that time planning a career in journalism) both principles and examples of the right way and the wrong way to work with journalists. He himself had begun as a journalist, and when I was in high school and at University I originally thought to follow in his footsteps, working three summers as an intern at a small but prosperous newspaper in my home state of Michigan.

For much of the content of this article, however, I am also indebted to a wonderful resource prepared by Dave Thomas of the American Geophysical Union, presented at an AGU meeting workshop many years ago, called *Communicating with the Media, Tips and Techniques.*

Finally, I also bring to this article more than twenty years' experience as the director of Public Relations at the Vatican Observatory. Most of the lessons I have learned in that role, I confess, come from mistakes that I have made over the years! I hope that by sharing these ideas and stories I will be able to reduce some of the anxiety that we can experience dealing with the press.

Before we go any further, however, there are some fundamental questions that each of us has to answer for ourselves. Ask yourself: How much of your time should be dedicated to public relations? To Press Relations? To Education and Public Outreach (EPO)? How much of your time can you commit to the never-ending demands of the news media? How much of your time can you commit to the general public through public lectures, and conducting tours of groups? When, and how, do we say "no"?

Each person will have a different answer to these questions, because each of us has a different set of skills, interests, and abilities when it comes to dealing with the press. There is no rule that says that every astronomer must also be a skilled interview subject. And even if handling these interviews is something we enjoy and feel good about doing, nonetheless each of us also has a full plate of other tasks that limit just how much we can do. We can't do everything.

That said, I wish to begin with a fact of life for astronomers at the Vatican Observatory that is special to us: every time we are quoted in the press, we will be identified as being representatives of The Vatican. Each of us will be labeled "The Pope's Astronomer" no matter how often we try to make it clear that we're just one of a team of astronomers. And no matter how often we might want to say that our opinions are merely our own, as astronomers who are well aware of our own fallibility, nonetheless any speculation we muse about in front of a reporter will inevitably be reported as "The Vatican believes…" We must always be aware of this. We are powerless to prevent this. But the Vatican connection is, after all, most likely the reason why the press is interested in speaking with us in the first place.

On the other hand, the fact that the press is interested, and wants to interview us, means that we in turn have certain rights and indeed a certain power to bring to the table. You always have the right, and the power, to ask what the interview is going to be about. You have the right to delay any interview until a time that is convenient to you. You can always turn down any interview request, for any reason, if something doesn't feel right to you about the situation. You also have the right to suggest they interview another person, if perhaps someone else would be a better fit for what they are looking for (and for your own crowded schedule).

You have the power, and indeed the responsibility, to negotiate about when, where, and how the interview will take place. Always negotiate!

Furthermore, as we mentioned above, you must always be aware that the interviewer will have an agenda that may well be different from yours. You have the power to not go where they want to drag you… either in terms of the topics to be covered, or the physical location where they want to interview you.

We put up with interviews because they are an important way that we can fulfill our mission to "show the world" about who we are and what we do. We need to be visible in the press. But a fundamental truth to remember is this: reporters need us as much as we need them. With out us, they have no story.

And we should also be sympathetic to the constraints that journalists work under. They are assigned stories by editors with little regard to their own interests and backgrounds. On any given day, the journalist may have to juggle half-a-dozen stories on topics that they have only the slightest familiarity with.

The baleful role of the journalist's editor can never be underestimated. When it comes to delivering a story, the journalist may want it *good*, but the editor wants it *now*. Likewise, headlines are never written by the journalist, but by someone who may completely miss the point of even the best-written story. As a result, every journalist knows that tried-and-true clichés are safe: pre-digested ideas go down easy.

But in fact, you can use these constraints to help the journalist tell a good story and get our message across. The first task on our part, before any interview, therefore is to have a clear idea in our own minds about what it is that we actually want to appear in the story. Start with finding out exactly what the journalist is going to ask about, and then work out well ahead of time what it is that you will want to tell them.

It's not by accident that articles in a newspaper or a television broadcast are referred to as "stories". And so when preparing what you want to say to an interviewer, it's important to think of your message in terms of the classical structure of a "story." In 1863 the German playwright Gustav Freytag, following ideas on literary analysis going back to Aristotle, outlined five stages in any typical story: exposition, development, climax, consequences, and dénouement.

A story starts with an exposition, a setting: Where are we? who are the protagonists? A place and people we care about… the poor orphan with the funny scar living under the stairs at Number 4, Privet Drive; the peaceful countryside with lovable hobbits.

A story lures you in with a problem, a conflict, a mystery; it makes you want to know what happens next, what happens when the wizard visits the orphan or the hobbits, and drags them all away from home.

A story has a central point, a climax, when the main character makes a crucial decision, when someone uncovers the key piece of information and everything you thought you knew is changed.

And the decision, the information, the main moment has consequences. It matters. We know how it matters, within the realm of the story, because we've been given the set up in the first place, back when the story began. So we know about the evil wizard who killed the hero's parents, we know about the evil ring that can control all the other rings of magic but corrupts one's soul in the process. But it also matters outside of the universe of the story; inside our own personal universe, it matters in a different way; it matters enough that we want to keep turning the pages to find out what happens next. We have to see how the decision plays out.

And then we come to the end, the resolution, the denouement. For a story to satisfy, you need to have a moment at the end to step back, catch your breath, and just take in the scenery… look over the landscape now, where Middle Earth will never be the same; where a new generation boards the train to Hogwarts.

So how do you apply this to preparing for an interview? Your "story" need the same parts. You describe the problem that you've been working on. You describe why it's a problem—and why you needed that clever or difficult or special thing you did, to make it all work. You describe your brilliant contribution. You describe what came of it. And then you sit back, see how it changes all our ideas about the origin and evolution of the universe, and why you'll be asking for another grant next year to keep up the good work.

This exercise automatically brings you to addressing questions you might otherwise overlook. For example, when describing the "setting" of your story, you need to know what it is you are assuming your audience already knows. And in order to do that, you have to know who your audience is. What level of education is the journalist aiming at? Are they writing for *Science* or the *Weekly World News*?

To know that, you have to do some research on the journalist themselves. What is their outlet? What other stories have they written? What sort of details do they like to include in their stories? In other words… before you even agree to the interview, do an internet search on the person requesting the interview.

If the journalists are coming with a film crew, find out the title and brief description of their program, the names of all of those who will be on the filming crew, and date and time of the shoot; and find out what their agenda might be! In this context, be aware that the major specialty cable networks like The History Channel or The Discovery Channel may have started out as attempts at quality television, but they are quality no more. The lure of large audiences from so-called "reality shows" has driven the better documentaries off those channels, with the result that if you are filmed for a show destined for cable TV you are likely to be seen alongside the worst sort of UFO or astrology rubbish.

As a result, one of the simple rules of thumb that I have adapted for the Vatican Observatory is that we do not accept film crews for any of the cable networks anymore; not Discovery, not History, not even National Geographic. It is too damaging to our reputation as a serious scientific institution to be included alongside the other programs they run.

Another rule of thumb is that if the web site of the reporter or program requesting an interview mentions aliens or UFOs in any form, no matter how seriously they claim to be, we will not allow them to interview us. Period. (This rule comes as the result of one of those learning-mistakes; we have been burned badly in the past by writers who claimed one thing while interviewing us, and then their program and book completely misrepresented what we had said.)

Finally… trust the smell test. If it smells funny, even if you can't put your finger on the problem, don't let it in your door.

(Of course, there are exceptions to all these rules. A journalist who has earned your trust, one with whom you have had previous good experience, can be trusted.)

Once you have agreed to do an interview, your work has only begun. First of all, all journalists must be made aware of the Vatican's rules for reporters coming for an interview (including filming or other recording) on Vatican territory. Reporters need to obtain written permission from the Vatican before they can record any sound, pictures or film on site. The details of how they can obtain these permissions may change over time, but the Vatican Observatory web site www.vaticanobservatory.va should have links to the current rules and addresses of those who need to give permission. While the rules are not quite as strict for recording in Arizona, there are University of Arizona regulation and Forest Service permits that must be cleared before filming can be done on campus or at the telescope on Mt. Graham.

Next, before the journalist arrives, prepare a very clear statement of what you want to say. Write it out, write it down, rehearse how you will say it! Anticipate the most likely questions; prepare an initial brief answer, a sound bite that's exactly 7 seconds long. Then prepare a longer answer, say 30 seconds, which you can use if you are given the time to expand on your story. Be prepared to illustrate your point with examples and analogies. Always restate the initial brief answer at the end.

During the interview itself, there are some useful points to remember. First and foremost, passion sells your story. If you're not excited about your result, your story, and you don't communicate that to the reporter (and ultimately to the audience) then they will have no idea that there's anything in what we are doing that is worth being passionate about.

There are also some simple rules that will make the journalist more likely to use your answers and less likely that you will make a terrible faux-pas in public.

Give your answers in complete sentences that make sense even without the question.

Don't say anything you wouldn't want to see on the front page of your country's biggest newspaper. Nothing is ever "off the record."

Don't repeat the reporter's framing of the question, especially if you want to refute it; use only your words to convey your ideas, not theirs.

Don't be tricked into saying more than you want to say by fear of "dead air." When you have said your piece, stop. Filling long pauses of "dead air" is their problem, not yours!

Avoid any temptation to be clever or sarcastic; it can backfire. No… rather, let us say, it *will* backfire. Always.

If you are being filmed, dress for the camera. Members of the Observatory should preferably be dressed in clerics for any formal interview. In any event, avoid clothing with stripes, as they can show up on video as odd Moiré patterns; and avoid material that makes noise when you move, because any such noise will be magnified by lapel microphones.

A lot of what the audience will remember about you comes not from what you say but how you look. Be aware of your body language. Keep your upper torso straight; head straight, shoulders down. Avoid looking defensive: don't cross your arms or clasp your hands, but hold your arms and hands out to signify you have nothing to hide. Use your hands for emphasis, within reason. And don't forget to make eye contact with interviewer!

Likewise, your face conveys a lot even when you are saying nothing. Smile when appropriate; astronomy is fun! Keep your mouth closed when listening. Be animated: open your face up, change your expression, let your eyebrows raise when appropriate. Don't lick your lips or bite your lip. And in general, don't touch your face with your hands.

But then… having come prepared to tell your story, what happens if the reporter never gets around to asking you about what you wanted to say? There's a trick to use, called "The Bridge". Listen to the reporter's question, acknowledge it, but then create a "bridge" to the answer you really want to give. That's when you can move to your previously prepared message.

If the questions are difficult, if the journalist is being aggressive or appears to be looking to trap you, be calm but don't follow their lead. Let your answers be driven by your message, not their questions. Don't repeat the "bait" word that the reporter suggests, but respond using only the words that you want to use to describe your ideas.

Whether hostile or note, there will always be questions that may throw you for a loop. Don't be afraid to say you don't know the answer to a question. If you can, refer the reporter to someone who does know, better than you. (And if you lose your train of thought… pause two seconds; tilt your head; act like you've just had a bright new idea. And then start talking about whatever you really wanted to talk about. Take advantage of the fact that we're astronomers—don't be afraid to look like the absent-minded professor!)

If you have bad news to report, certain basic rules can help keep your interview from making the matters even worse. Don't respond with hostility or emotion. If you can't answer, give a reason why. Never say "no comment"; if you say "no comment" you actually wind up conveying the idea that the premise of the hostile question is true but you aren't prepared to admit it in public!

Keep personal opinions to yourself. Never lie. Never guess or speculate; especially, never speculate about the motives of people who disagree with you.

And one of the hardest rules to keep, but one of the most important: never answer a "hypothetical question." We cannot say what we will do, if… aliens are discovered, or the telescope crashes, or terrible poisons from space are emitted from a freshly fallen meteorite. To answer such questions, first of all, dignifies the often absurd propositions; and furthermore, the way we will respond in any such event will depend so much on the details of the situation that can't possibly be known ahead of time. You do not want to go on record saying that you would do or think one thing, only to have to backtrack when the reality of a situation shows that a completely different response is more appropriate.

The work does not stop once the interview is over. Learn from your experience: what worked? What didn't?

Getting “always the same questions” is OK: there’s comfort in knowing what is likely to be coming, what sort of answers work, how to pace the story, what details can be skipped over, where the laugh lines are.

Most scientists who are not used to being interviewed will inevitably have a feeling of exhaustion, let-down, or concern that things may not have gone the way you like. The biggest (but hardest) thing to do after the interview is over, is to forget about. At the end of the day, don’t worry if things go wrong. Don’t make a big deal of minor errors; corrections often just spread the error further. If there is a major error, you can contact the reporter; but whatever you do, don’t contact their boss!

Remember, at the end of the day, there is no such thing as bad publicity. Anything that creates awareness of astronomy or the Specola is good. Indeed, more often than not, the reaction of most people even to bad publicity is “gee, I didn’t know the Vatican had an observatory!”

And that, in turn, reminds us why it is worth the headache of making ourselves available to the press. It can be a lot of fun! Your family and friends get to see your name in the paper! And as the result of such interviews, I’ve gotten to meet a number of interesting (and famous) people, such as the science fiction writers Terry Pratchett and Gene Wolfe; the Astronomer Royal, Sir Martin Rees; and the astronauts Nicole Stott and Roberto Vittori. More people ask me about meeting Stephen Colbert than about meeting the Pope!

And in fact, this sort of publicity has garnered real benefits to the Specola up to now. Many email fan letters, from folks who first learn about us and the Church’s support of science, arise from our interviews. They can often be heartening stories of young people who were afraid a life of science might lead them away from God, or from older scientists who just like to be reminded that they are not alone in their lives of faith.

And in fact, it was a *Sky and Telescope* article that led to Rich Friedrich joining the Vatican Observatory Foundation; for many years he has served as chairman of our board.

Popularizing our science is just as much work as writing a scientific paper. It takes a special set of skills. If you aren’t good at it, admit it: ask for help. Give help to others who ask.

And it’s important. We astronomers are here to feed a common human hunger to know about ourselves and our place in the universe. The Church supports us not merely to show the world that she supports science, but because studying creation is a sure route to getting to know the Creator. It is our job as astronomers and Jesuits to show the whole world where one can find God in all things, on Earth and in the heavens.

# The Rome Historical Cauchoix Telescope Recovered

**Aldo Altamore, Francesco Poppi and Tommaso Bosco**

The achromatic refractor Cauchoix was acquired by the Astronomical Observatory of Collegio Romano in the year 1825. In the second half of the Nineteenth Century, Angelo Secchi used this 16 cm aperture telescope, coupled with the objective prism, to perform the observations on which his pioneering spectral classification of stars was based Assembled.

on a new equatorial mount in 1885, the instrument has been in service at the Astronomical Observatory of Rome until 1980 and then dismantled.

Recently, we found the achromatic doublet of this historical instrument, which had gone missing for more than 30 years.

The finding will allow to reconstruct the configuration of the telescope as it was at Secchi's time.

## Introduction

The second half of the 19th century, with the formalisation of the laws of electromagnetism, was marked by the culmination of classical physics. Simultaneously, experiments were carried out which soon led to the new scientific revolution of quantum mechanics and relativity.

This era was also characterised by a rapid evolution of applications of scientific knowledge and by the introduction of new technologies both in daily life and in the

A. Altamore (✉) · T. Bosco
Mathematics and Physics Department, Roma Tre University, Rome, Italy
e-mail: altamore@fis.uniroma3.it

A. Altamore
Vatican Observatory, Albano Laziale, Italy

F. Poppi
Bologna University, Bologna, Italy
e-mail: francesco.poppi@gmail.com



G. Gionti, S.J. and J.-B. Kikwaya Eluo, S.J. (eds.), *The Vatican Observatory, Castel Gandolfo: 80th Anniversary Celebration*, Astrophysics and Space Science Proceedings 51, https://doi.org/10.1007/978-3-319-67205-2_13

research field. It is worth underlining the progress made in the fields of energy resources and materials, telecommunication and transport, machine tools and precision instruments.

In the field of astronomy, a particularly important role was played by photography which, for the first time, permitted to obtain images of celestial objects through a technique alternative to traditional scientific drawing.

This extremely lively scientific and technological climate led to what was described by Samuel P. Langley (1834–1906) as the *new astronomy* or *physical astronomy,* otherwise known as astrophysics.

The emergence of astrophysics was the direct consequence of the pioneering research of Gustav Kirchhoff (1824–1887) and Robert Bunsen (1811–1899) who, during the 1850s, had paved the way for the investigation of the physical and chemical properties of materials. Their research was mainly carried out through laboratory analyses of the light emitted or absorbed by various chemical elements using the spectroscope, invented several decades before by the German physicist Joseph von Fraunhofer (1787–1826).

The new science grew within a very different context from that of traditional astronomy.

In fact, astronomy was at that time restricted to the compilation of catalogues of celestial bodies, to the calculation and prediction of the motion of celestial bodies through the Newtonian physics, as well as to the morphological studies of the surfaces of planets and of extended images of object such as nebulae and star clusters.

In order to appreciate the extent of the change, it is worth recalling that Auguste Comte (1798–1857), the founder of Positivism, had categorically excluded the possibility of measuring the temperature and chemical composition of celestial bodies in *Cours de la Philosophie Positive* published in 1835.

However, only a few decades after Comte's assertion, the Jesuit Angelo Secchi (1818–1878) developed the first spectral classification of the stars which made it possible to evaluate their physical and chemical properties.

Secchi became director of the Collegio Romano Observatory in 1849, succeeding to Francesco De Vico S.J. (1805–1848). Starting from 1852, he began a major reorganisation of the Observatory by building on the roof of the church of Sant'Ignazio what can rightly be regarded as the first astrophysical observatory in the world, Fig. 1. In the new observatory two refracting telescopes were installed (Altamore 2011):

- the Cauchoix achromatic telescope, which was characterised by a 16.9 cm aperture and 238 cm focal length, which was in the dotation of Collegio Romano Observatory since 1824.
- the Merz refractor (∅ = 24,4 cm, f = 433 cm) which was purchased by Secchi in 1854 thanks to the financial support of his colleague Paolo Rosa S.J. (1825–1874).

These two telescopes had a different story. The Merz remained in operation at Collegio Romano Observatory until 1889, year in which a new equatorial mounting was installed on the granite pillar to support the new refractor Steinheil-Cavignato

**Fig. 1** The Collegio Romano Observatory in the beginning of 20th century (OAR-INAF Archive)

(Tacchini 1901). The director Pietro Tacchini (1838–1905) decided to place the telescope in the collection of the Copernican and Astronomical Museum, which had been established in 1886 thanks to the Polish historian Arturo Wolinsky and Tacchini himself (Calisi 2000).

The Merz remained in the museum exhibition until the mid 50's of the XX century when Giuseppe Armellini decided to put it back in operation and placed it alongside the Steinheil-Cavignato, which since 1938 was operating in the main dome of Villa Mellini, venue of the Astronomical Observatory of Rome. This choice was fatal to the instrument. On the night between the 15th and the 16th of July 1958, the telescopes Steinheil-Cavignato and Merz were destroyed by a fire.

The Cauchoix telescope remained in service for al longer time, assembled on a new equatorial mount in 1885, the instrument has been in service until 1980. In the following sections we will try to illustrate its long life.

## The Cauchoix Telescope's Long Story

This telescope is a key instrument in the history of stellar and solar astrophysics.

Acquired in 1824, it was installed on a quite sophisticated altazimuth mount shown in Fig. 2.

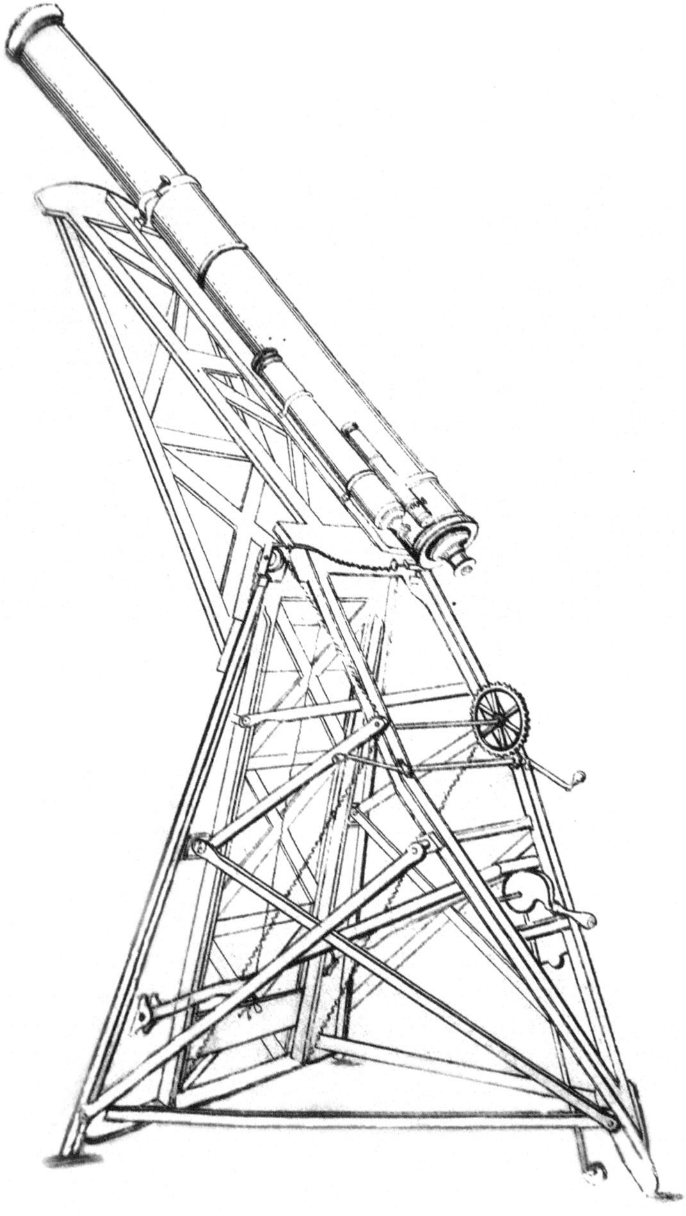

**Fig. 2** An engraving showing the original altazimuth mount of Cauchoix Telescope (1824) (Vatican Observatory Archive)

When Angelo Secchi (1818–1878) renewed the Collegio Romano Observatory, he installed the telescope on the equatorial mount which today is exposed in the astronomical Museum at Monte Porzio Catone (Fig. 3).

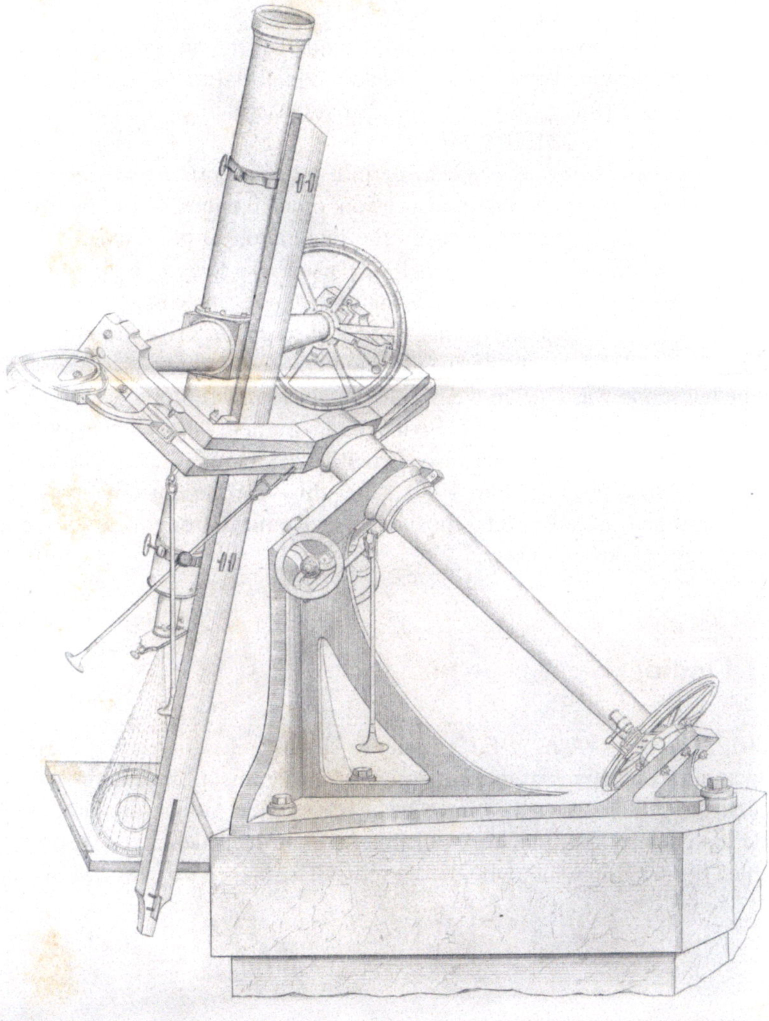


**Fig. 3** An engraving showing the equatorial mount of Cauchoix Telescope acquired by Secchi in 1854 (Vatican Observatory Archive)

The telescope was mainly used for solar observations. Indeed in the second half of 19th century, the investigation on the physical and chemical properties of the Sun was one of the main research lines of the newly born astrophysics. Secchi gave major contributions to the physical understanding of phenomena such as spots, flares and prominences respectively taking place in the solar photosphere and chromosphere. The observations were registered either with the traditional drawing technique, or by photography.

In addition, by adding the objective prism to this instrument, Secchi collected a great deal of observations on which he based his first star spectral classification, on which stellar astrophysics lies upon.

In 1886, owing to its exceptional optical quality, the objective was mounted on a new tube and provided with a more modern equatorial mounting built by Giuseppe Cavignato (masterful technician at the Padua Observatory) according to the project proposed by Tacchini (1901), Fig. 4.

The mounting was conceived to permit the transportation of the instrument in different sites, especially on the occasion of campaigns for total solar eclipses. Indeed it is possible to change easily the inclination of polar axis.

In 1913 Giorgio Abetti (1882–1982) used the telescope to obtain a set of objective prism spectra plates of a sample of the stars used by Secchi for his classification, in order to verify his work and to characterize the optical properties of the prism utilized by the Jesuit astronomer.

The telescope was later moved to the newly founded Monte Mario Observatory. The last research work performed with the instrument was in occasion of the 1970 transit of Mercury on the solar disc (Croce 1973). However, the Cauchoix was still used for didactic purpose until 1980. After this year the objective was definitively dismounted and placed in the repositories of Rome Astronomical Observatory and forgotten for about 30 years.

## The Finding

In 2012, two of us (A.A. & F.P.) began to chase the lost objective, starting with the collection of objectives of the Copernican Astronomical Museum in Monte Mario.[1] At first the search seemed to have been successful. An achromatic doublet of the same diameter as the Cauchoix objective was in fact identified. However our colleague Giorgio Buonvino, who before retiring had taken care for and maintained the

[1]The Astronomical and Copernican Museum, created in 1873–77 by the Polish scholar Arturo Wolynski (1884–1893), includes two sections today: that of Monte Mario (Villa Mellini) which houses mainly the oldest instruments, and that of Monte Porzio where the XIX and XX century instruments are kept.

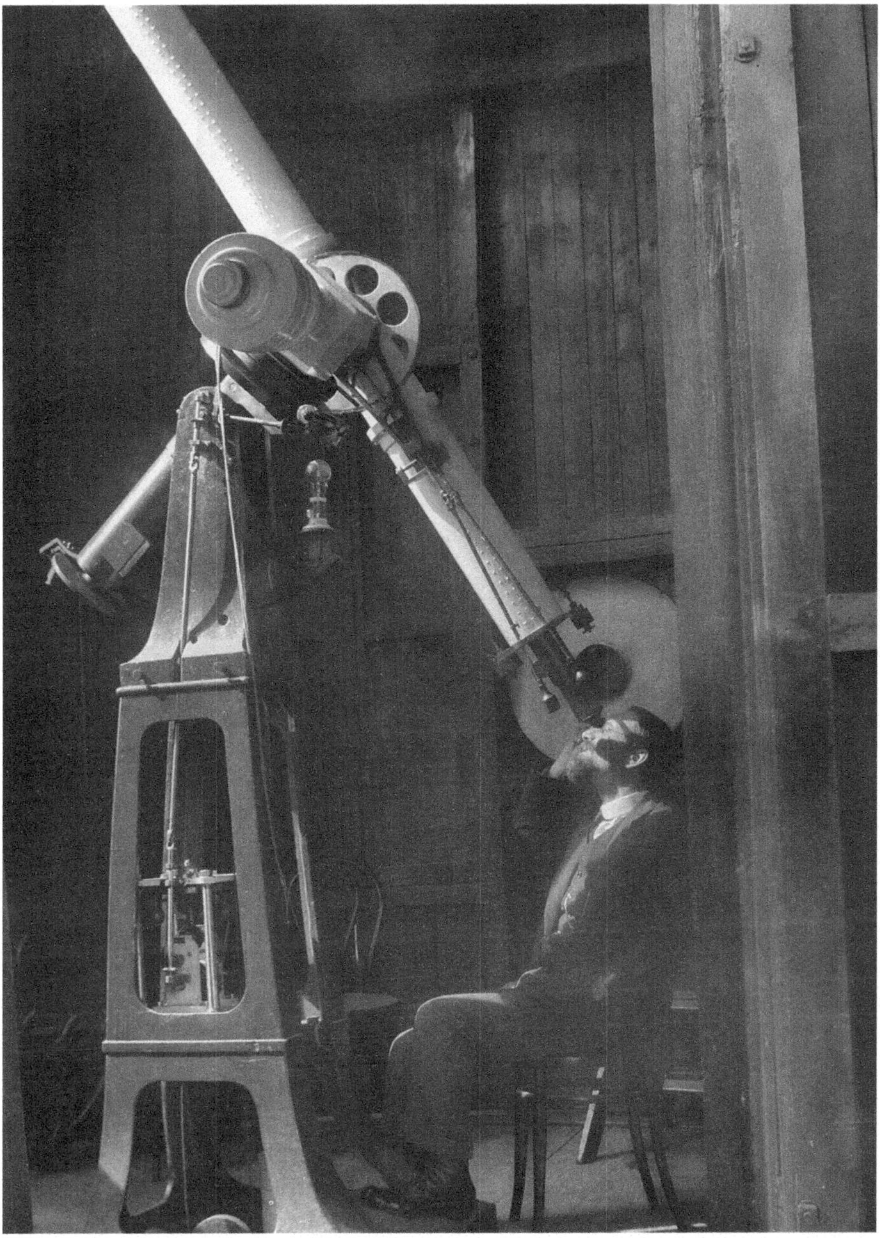

**Fig. 4** The Cauchoix Telescope on equatorial mount acquired by Tacchini. The photograph was shot in the dome of Collegio Romano Observatory at the beginning of 20th century (OAR-INAF Archive)

**Fig. 5** The cookies box which was used to store the objective of Cauchoix for about 30 years

Rome Astronomical Observatory instrumentation, was sure that this was not the right objective. Also a rough verification led to the conclusion that the focal length of this lens was about twice that of the Cauchoix objective used by Secchi. Subsequent researches has shown that the objective belonged to the Horizontal Solar House Askania, another historical instrument which will be discussed in a later section. Buonvino himself told us that when the instrument was dismantled, he placed the objective in a metal box of those usually used for cookies. Our research was therefore put on hold until after a few months the cookies box was accidentally rediscovered in Monte Mario, in the warehouse which houses objects not on display in the Museum. The box, shown in Fig. 5, was still sealed as when it was placed in the repository thirty years earlier. A label with the words "Cauchoix objective" was pasted on it, the lens was kept in it, wrapped in computer printouts of the era in which it was placed.

It was not however possible to find the tube of the telescope, the most probable hypothesis being that it was thrown away around the year 2000, on the occasion of the renovation of the premises in which it was kept.

## The Reallocation of the Objective

Taking into account that it was no longer possible to re-install the instrument on the late nineteenth century mount, it was decided to investigate the possibility to mount it on the tube of 1824 on display in Monte Porzio Catone. To the surprise and satisfaction, this operation was extremely easy because the ring with its thread turned out to be the original one.

Figure 6 shows the moment when the objective was remounted on the old tube. Immediately after installation of the lens, we verified that the objective prism might be mounted; also this operation was a total success. So today the optical configuration of the ancient tool, including the objective prism, is completely restored.

## The Horizontal Solar Camera Askania

This instrument too has great historical significance as it is the only surviving example of the two crafted in the early decades of the 900 by the company Askania Bamberg.

It is a horizontal refracting telescope with an objective of 16 cm diameter and a focal length of 4.8 m, equipped with two coelostats of 30 cm diameter and a motor for tracking. The images of the Sun were collected on 13 × 18 cm photographic plates.

The first Horizontal Solar Camera Askania was built by the well known German company on behalf of the Astrophysical Observatory of Potsdam, on the occasion of the total solar eclipse of 1914 (Wolfschmidt 2011). This instrument was sent to the Crimea at the time of the eclipse, and was lost as a result of events subsequent to the outbreak of the First World War.

The second sample was installed at the Campidoglio Observatory in 1924, Fig. 7. In 1938, with the founding of the Astronomical Observatory of Rome, it was transferred to Villa Mellini on Monte Mario.

In the postwar years, the Camera was placed in a special pavilion, built at the base of the Solar Tower, which was later demolished to build the round building that now houses the CED-INAF.[2] It remained operational until the mid-60s for the daily observations of the solar photosphere and for the observations of solar eclipses, during several expeditions in Italy and abroad (Cimino 1965). Figure 8 shows the instrument installed in Imperia during the total solar eclipse of February 15, 1961.

[2]This pavilion, designed by the architect Saverio Busiri Vici, was originally intended to host the electronics laboratory, www.saveriobusirivici.it/#4.

**Fig. 6** The moment of reallocation of the objective on the old tube

**Fig. 7** The Askania Horizontal Camera placed in dome of Senatorial Palace of the Campidoglio, which housed the Ertel Meridian Circle (OAR-INAF Archive)

**Fig. 8** The Askania Solar Camera installed in Imperia during the total solar eclipse of February 15, 1961 (OAR-INAF Archive)

At the OAR-INAF, in addition to the instrument, many thousands of plates produced by it in the course of more than forty years of operation are now stored.

## Conclusions

The heritage of historical scientific instruments in Italy is immense and invaluable; regarding particularly the astronomical instruments, it is believed that the value of the collections held by INAF is equivalent to that of scientific equipment currently in use, including large telescopes operated by the institution.

It should also be remembered that in addition to the collections kept by large institutions, a widespread heritage is present in Italy in small structures, considering the historical instruments that are part of the laboratory equipment of schools. In some cases these collections include tools and objects that date back to the 17th and 18th century.

The enhancement of this heritage of ancient instruments is of great importance, not only for the history of science but also because these testimonies are an effective tool for the dissemination of scientific knowledge among the public, and especially among young people for training new generations of scientists.

## References

Altamore, A. (2011). in Altamore e Maffeo, edit., *Angelo Secchi- L'Avventura scientifica del Collegio Romano*, p. 120.

Calisi, M. (2000). *Storia e Strumenti del Museo Astronomico e Copernicano di Roma*, OAR-INAF.

Cimino, M. (1965). Rapporto Annuale OAR, p. 51.

Croce, V. (1973). Icarus, 20, 179.

Tacchini, P. (1901). Mem. R. Oss. Coll. Romano, Ser. III, v. I.

Wolfschmidt, G. (2011). *Telescopes made in Berlin: from Carl Bamberg to Askania,*in *From Earth-Bound to Satellite*, p. 177.

# The Vatican Meeting on Stellar Populations

José G. Funes, S.J.

## The Vatican Observatory's Mission

The Vatican Observatory is a unique research institute. The Catholic Church has always been aware of the importance of astronomy in human culture and has embraced, encouraged and promoted it. The Vatican Observatory is a concrete sign of that commitment. Its mission is to be the presence of the Church in the world of Sciences, particularly in Astronomy. Vatican astronomers are called to serve the Pope and their colleagues, doing astronomical research and helping others to collaborate scientifically through research projects and the organization of schools and scientific meetings.

Astronomy is a wonderful science and still a very exciting frontier to be at. Vatican astronomers hope to serve humanity in asking the deepest human questions about the origin of the Universe and its philosophical and theological implications. In accomplishing this mission, the staff is close to the Pope—even physically, for decades the headquarters have been at the papal residence of Castel Gandolfo—on one side, and on the other side Vatican astronomers are at the border with the wider world and therefore open to dialogue with everyone, with those who believe and those who do not.

A good historical example of being close to the Pope and, at the same time, at crossroads of science, history, and culture is the relationship that Father Daniel O'Connell had with Pope Pius XII. O'Connell directed the Vatican Observatory (1952–1970) and was President of the Pontifical Academy of Sciences (1968–1972) succeeding in that position Msgr George Lemaître, the "Father of the Big Bang cosmology".

O'Connell wrote on Pius XII and the Sciences: "I can, however, say something about his discourses on scientific topics. Here again he has been criticized for going

J. G. Funes, S.J. (✉)
Vatican Observatory, Albano Laziale, Italy
e-mail: josegfunessj@gmail.com



G. Gionti, S.J. and J.-B. Kikwaya Eluo, S.J. (eds.), *The Vatican Observatory, Castel Gandolfo: 80th Anniversary Celebration*, Astrophysics and Space Science Proceedings 51, https://doi.org/10.1007/978-3-319-67205-2_14

into too much technical detail in these addresses, as if he were trying to teach the experts their own business. That was certainly not his aim. He wished his listeners to feel that he was interested in their problems and sympathized with their difficulties. Then he invariably used this technical introduction to lead up to religious and moral considerations appropriate to the subject and to the audience."[1]

O'Connell, a privileged witness, continued: "In parenthesis it might be well to correct certain misapprehensions. Many have wondered how the Pope could speak so learnedly on such a great variety of topics. Some indeed have asserted that he prepared all his addresses himself, alone and unaided; others have even gone so far as to suggest that all this knowledge was infused by special divine inspiration. Both notions are incorrect. Pope Pius XII was indeed an ardent student all his life (he had, for instance, a life-long interest in astronomy), but he was far too conscientious and too intelligent to attempt to pose as an authority de *omni re scibili*. He was fully conscious of the obligation, imposed on him by his high position, of weighing his every word. This reinforced his natural passion for exactness and precision, so that he took every care to avoid making any kind of incorrect statement. Thus, when necessary, he called to his aid an expert (sometimes more than one) in the field he intended to discuss. He studied very carefully the technical material presented to him and made it his own, and molded it so as to drive home the religious or ethical considerations he wished to put before his audience."[2]

As an example, O'Connell referred to the Pope's last address on Astronomy delivered on 20 May 1957, when inaugurating the Study Week on Stellar Populations organized by the Pontifical Academy of Sciences and the Vatican Observatory.[3]

But before considering the importance of the Vatican conference on *Stellar Populations*, it is necessary to provide a historical framework of this meeting. It is good to keep in mind that Astronomy and Cosmology were very much cutting-edge disciplines in the 50s and 60s. Also this was the time of the Cold War and of the big debate between the supporters of the Big Bang and Steady State theories.

## Religion, Politics, and the Universe

This is the title of the section that Kragh[4] dedicates to the XXX in his book *Cosmology and Controversy*. Kragh provides a historical background of two themes that are present in post-1920 cosmology: the stationary and evolutionary

---

[1]O'Connell (1958).

[2]Ibid.

[3]*Stellar Populations*, 1958, Proceedings of a Conference Sponsored by the Pontifical Academy of Sciences and the Vatican Observatory held at The Pontifical Academy of Sciences on 20–28 May 1957, edited by D.J.K. O'Connell.

[4]Kragh (1996), p. 251.

universe represented by the Steady-State theory and the Big-Bang theory, respectively.[5]

In the 1940s, George Gamow, a nuclear physicist, proposed the Big Bang theory. The works of Alexander Friedmann and George Lemaître prepared the road to this theory in the 1920s.

Sir Fred Hoyle was the main proposer of the steady-state theory, formulated in 1948 by Bondi and Gold, a model of dynamic equilibrium involving the continuous creation of matter throughout the universe. He was the most vocal and most intransigent critic of the Big Bang cosmology. Hoyle is one example of the potential relevance of anti-religious views in the choice of cosmological models.[6] In 1952 O'Connell strongly criticized Hoyle: "The fact is that, though Hoyle sets out expressly to teach philosophers and theologians their business, he makes no serious attempt to find what they hold, or what reasons they give for their beliefs."[7]

The controversy between the two competing theories was developed in the 1950s and 1960s and was decided in favor of an evolutionary universe by the observation of the cosmic microwave background radiation in 1965 by Penzias and Wilson.

As McMullin pointed out,[8] it was the first time that a physical science was led to assert the beginning of time and therefore to estimate the age of the universe. The potential relevance of this finding was perceived by believers and non-believers.

As Kragh points out Pius XII was a learned and enlightened man with an interest in astronomy and other sciences. He was fascinated by the theory of the expanding Universe.[9] On 22 November 1951 the Pope addressed the Pontifical Academy of Sciences endorsing the Big Bang picture: "everything seems to indicate that the material universe has had, in finite time, a powerful start, provided as it was with an unimaginable abundance of reserves in energy; then, with increasing slowness, it has evolved to its present state."[10]

The address had an important impact on the scientific community. The Physical Review in 1952 published a paper by Gamow introduced by a lengthy quotation of the address by Pius XII.[11] The endorsement of the Big Bang cosmology by the Pope when the theory was not unanimously accepted provoked strong criticisms accusing him of using these scientific ideas to his purpose of proving concordance between science and religion.[12] Even Lemaître was unhappy with the unqualified use of the hypothesis in this context. On a footnote of his chapter *How should cosmology relate to theology*? Mc Mullin offered his personal testimony as he was

[5]Ibid., Preface.

[6]McMullin (1986), pp. 32 and 34.

[7]Kragh (1996), p. 195.

[8]McMullin (1986), p. 30.

[9]Kragh (1996), p. 256.

[10]http://inters.org/pius-xii-speech-1952-proofs-god, accessed on 27 October 2017.

[11]Kragh (1996), p. 256.

[12]Ibid. pp. 257–258 and McMullin (1986), p. 32.

attending a graduate seminar with Lemaître in 1951 on his return from the Academy meeting in Rome.[13]

As Tanzella-Nitti notes, it is probable that one of the academicians, Edmund Whittaker, an English physicist-mathematician already advanced in years and interested in the relationship between science and faith, prepared part of the draft of the speech[14]; while McMullin mentioned that at the time it was said that the principal author was Father Agostino Gemelli, President of the Pontifical Academy of Sciences.[15]

It is worth noting, that Gamow sent to Pope Pius XII a popular article in 1951 and later his book on *The Creation of the Universe* (1952) through the apostolic delegate. According to the apostolic delegate the "Holy Father read it with satisfaction and looks forward to the publication of your book on the Creation of the Universe". Gamow was not Catholic or had any special relations with the Catholic Church but he knew that Pius XII had endorsed the Big-Bang theory in his address of 1951 to the Pontifical Academy of Sciences.[16]

Tanzella-Nitti concludes that, thanks to the intervention of O'Connell and Msgr. Angelo Dell'Acqua who worked for the Secretary of State, it is plausible that Georges Lemaître spoke with Pius XII regarding the November 1951 speech and that he offered the Pontiff some clarifications and some suggestions the text of the speech held at Castel Gandolfo on September 7, 1952 before the members of the International Astronomical Union. Although, in this address the Pope avoided any specific references and religious implications of the Big Bang theory, the New York Times titled in its edition of 8 September 1952: "Pope says science proves God exists."[17]

As Kragh points out, the International Astronomical Union meeting in Rome illustrated the politically and ideologically sensitive aspects of cosmology. The four Soviet delegates stayed away from the discourse and the subsequent audience.[18]

## The Vatican Study Week on Stellar Populations

The concept of *Stellar Populations* is present in almost every introductory book to Astronomy. As O'Connell points out in the in the introduction of the proceedings: "the Problem of Stellar Populations, first suggested to me by Dr. J.J. Nassau, proved to be eminently suitable. Ever since Baade, in his 1944 paper, distinguished two types of stellar populations, the subject has aroused great interest among

---

[13]McMullin (1986), footnote 25, p. 53.
[14]Tanzella-Nitti (2016).
[15]McMullin (1986), footnote 25, p. 54.
[16]Kragh (1996), p. 117.
[17]Ibid., footnote 187, p. 431.
[18]Ibid., p. 259.

astronomers and has led to a flood of researches in various branches of astronomy. Nevertheless, up to that time it had not been the subject of any conference or symposium. The time now seemed just right for a thorough discussion of all the problems involved."[19]

O'Connell continues describing the process of the organization of the conference —which has not changed much since those times except for the speed in communication—and describes the criteria for the invitation to participate: "The scientists to be invited were chosen with a view to covering adequately the various aspects of the subject, both in the formal papers and in the discussions."[20] This is especially significant for the case of the invitation to Hoyle who had an important role in the elaboration of the summary of the conference. It is easy to understand that, being Hoyle a strong anti-religious voice, there was some resistance in the Vatican and among pontifical academicians to invite Hoyle to the conference. As we will see the Pope himself intervened to approve Hoyle's participation.

Regarding the summaries and their discussion, O'Connell reported: "The Committee felt that it would be a great advantage to prepare a summary of the work of the Semaine (Week) that could be submitted to the members before the end of the meeting, so as to clarity ideas and bring the many different approaches to the central theme into relation with each other, and also to facilitate the drawing up of conclusions and recommendations. Mr. Hoyle and Professor Oort kindly consented to undertake this by no means easy task, the former from the physical, and the latter from the astronomical, point of view."[21]

It is worth noting that Pius XII was personally involved in the organization of the meeting. This results from my personal research in the Archives of the Secretary of State and of the Vatican Observatory. In a note of 10 November 1956, as an answer to a *pro memoria* by O'Connell to Pius XII, he should be informed that it is the Pope's desire that O'Connell be the President of the Study Week.

From another note of the archive of the Secretary of State, from an audience with Pius XII, it results that O'Connell received the following oral communication:

> "Communicated by voice to Father O'Connell *Ex Audientia* with His Holiness 27 February 1957
>
> (1) Considered what Father O'Connell affirms, Professor Hoyle can be invited.
> (2) The study-week can be prepared, like the others organized by the Pontifical Academy, with the collaboration of the Vatican Observatory and following as much as possible the budget of Father O'Connell.
> (3) Father O'Connell should meet Professor Armellini and explain the Hoyle case."

[19] D.J.K. O'Connell, "*Introduction*", in *Stellar Populations*, 1958, Proceedings of a Conference Sponsored By the Pontifical Academy of Sciences and the Vatican Observatory held at The Pontifical Academy of Sciences on 20–28 May 1957, edited by D.J.K. O'Connell, XI.

[20] Ibid. p. XII.

[21] Ibid. p. XIII.

In order to understand the importance of the Vatican Study week, I copy here the names of the participants that have made the history of Astronomy in the twentieth century:

> Giuseppe Armellini, Director of the Observatory of Rome, Walter Baade, Mount Wilson and Palomar Observatories, Adrian Blaauw, Yerkes Observatory, Hermann Alexander Brück, Director of Dunsink Observatory, Daniel Chalonge, Paris Observatory, William A. Fowler, California Institute of Technology, Otto Heckmann, University of Hamburg. George H. Herbig, Lick Observatory, Mr. Fred Hoyle, St. John's College, Cambridge, Josef Junkes S.J., Vatican Observatory, Georges Lemaître, Catholic University of Louvain, Berth. Lindblad, Director « of the Observatory of Stockholm, William W. Morgan, Yerkes Observatory, Jason J. Nassau, Director of the Warney and Swasey Observatory, Daniel O'Connell S.J., Director of the Vatican Observatory.[22]

The list provides a good picture of the importance, prestige and internationality of the Study Week. The Pope was informed by O'Connell of the success of the meeting thanking him for the papal address to the participants. O'Connell also expressed the common satisfaction as well as the serenity and the freedom with which the participants were able to accomplish the work.

I have just wanted to give an example of how the Vatican Observatory has been at the crossroads of science and religion, and no matter what was the religious or anti-religious position of their colleagues, Vatican astronomers have sought to collaborate with the best possible scientists to explore the universe.

**Acknowledgements** I am deeply grateful to H.E. Msgr. Giovanni Angelo Becciu, Substitute for General Affairs to the Secretary of State, for allowing me access to the Archive of the Secretary of State to consult the documents related to the Study Week on Stellar Populations. I am also thankful for the kind assistance of the staff of the Archive of the Secretary of State and to Father Maffeo, S. J., Mr. Federico Balzoni, and Mr. Antonio Coretti for assisting with collecting the numerous documents in the Archive of the Vatican Observatory.

## References

Kragh, H. (1996). *Cosmology and controversy: The historical development of two theories of the universe*. Princeton University Press.

McMullin, E. (1986). *How should cosmology relate to theology*? In A.R. Peacocke (Ed.), The sciences and theology in the twentieth century. University of Notre Dame Press.

O'Connell, D. J. K. (1958). Pius XII, recollections and impressions, studies. *An Irish Quaterly Review* (Winter), 362.

Tanzella-Nitti, G. (2016). The Pius XII—Lemaître Affair (1951-1952) on Big Bang and Creation. http://inters.org/pio-xii-lemaitre. Accessed on October 28, 2016.

[22]Ibid. pp. XI–XX.

# The Leap Second Debate and the Lessons from Timekeeping History

**Paul Gabor**

**Abstract** Timekeeping, together with navigation, is one of the oldest missions of astronomy, originating long before the dawn of written history. The ancient and sacred task of timekeeping, linking the eternal with the everyday, stood at the origin of the Vatican Observatory. The history of timekeeping schemes throughout millennia has been a search for a balance between the practical and the symbolic. The current debate focuses on the Leap Second, a mechanism first applied in 1972. Atomic clocks, careful observations of distant quasars, etc. tell us that the Earth spins somewhat irregularly. The symbolic value of coupling civil timekeeping to the heavens is so strong, however, that the international community still requires the International Telecommunication Union to broadcast time signals following Earth's rotation. In order to reconcile this symbolically significant requirement with the practicality of uniformly flowing atomic time, a leap second is added (or subtracted) from the otherwise perfectly regular sequence, so that every so often there is a minute with 61 (or 59) seconds. This system is currently under review, with a proposal on the table to abandon the leap second, decoupling civil timekeeping from astronomical phenomena. Social mechanisms involved in timekeeping can be studied by examining the historical evidence related to calenders. These lessons may provide a broader context for the leap-second debate and they may shed light on it. Calculated timekeeping does not actually agree with the heavens at any given moment, but this does not seem to jeopardize the general perception of its astronomical conformity since it is rightly perceived as a way of preserving it on average. The current proposal, however, clearly entails decoupling civil time from Earths rotation. If presented as a definitive solution, it would result in a direct and unprecedented breach with the principle of astronomical conformity. If presented as a temporary measure, to be applied until a better solution is found, it would preserve the principle of astronomical conformity because the latter fulfills its social function even when it is not observed perfectly considering that in the realm of symbolism what counts is the general perception, not the fact.

---

P. Gabor (✉)
Vatican Observatory, Tucson, AZ 85721, USA
e-mail: pgabor@specola.va

 
G. Gionti, S.J. and J.-B. Kikwaya Eluo, S.J. (eds.), *The Vatican Observatory, Castel Gandolfo: 80th Anniversary Celebration*, Astrophysics and Space Science Proceedings 51, https://doi.org/10.1007/978-3-319-67205-2_15

## Introduction

Contemporary society perceives reality in a partially new manner, somewhat informed by science. Many symbols, however, were anchored in another perception of reality. As a result, some symbols may be perceived as decoupled from this new world view. As Tillich puts it, true symbols participate in what they symbolize. When protesters burn a flag they are not merely destroying a piece of fabric. Furthermore, symbols "grow and die like living beings" (Tillich 1964). In Jungian terms, symbols inhabit the collective unconscious. An artist may propose many images but only a few will come to life as symbols.

The leap second debate has a substratum where science and symbolism clash. When Dennis D. McCarthy and William J. Klepczynski suggested abolishing the leap second (McCarthy and Klepczynski 1999), they hoped it would be straightforward. Their arguments were sound and aesthetically pleasing to all metrologists. And yet, the proposal has met with staunch opposition. In fact, the latest decision (World Radiocommunication Conference 2015) is to reconsider the matter in 2023.

The leap second may be a somewhat obscure issue. It certainly is very technical. And yet, it is clear that there is more to it than meets the eye. We shall first introduce the leap second, the proposal(s) for its redefinition, and the discussion so far. Then we shall describe some of the underlying factors, which have shaped the long history of calendars as the most prominent example of complex timekeeping schemes. And we shall conclude with a suggestion that could improve the discussion.

## The Leap Second Saga

### *Second: Solar, Ephemeris, and Atomic*

Plato's *Timæus* describes time as "a moving image of eternity," (37cd, Jowett 1892) identifying it with the diurnal, and annual motions of heavenly spheres. What Plato elaborates is not his own imposition or invention. He merely codifies what human societies have observed and perceived since "time immemorial" in the ritual nexus between time and eternity. Since the dawn of human race, we have relied on the uniform apparent motion of the Sun and other astronomical phenomena because they were the epitome of regularity. From a practical point of view, they also provided a very accurate and readily available source of time. The accuracy derived from Earth's rotation suffers from a fractional uncertainty of roughly $10^{-8}$ over an averaging time of $10^{7}$ s. Already Halley noticed an apparent acceleration of the Moon (1695) and Kant suggested that it was due to Earth's deceleration (1754) (Nelson et al. 2001). By the 1930s, astronomers also realized that the deceleration was not uniform.

The *second* was originally conceived as *the 1/86,400 of a mean solar day*. It became apparent, however, that the length of the mean solar day depends on the length of the averaging window and it evolves with time. This discovery placed the international community before a choice. Keeping the second linked to the (averaged) rotation of the Earth would have meant that the length of the second would not be fixed. A uniform time scale is indispensable for metrological purposes. A changeable second would not be suitable as the basis of such a time scale. Another unit would be required to construct a uniform time scale. The international community, however, chose the more natural option of fixing the duration of the second, decoupling it from Earth's rotation.

Between 1956 and 1958, the *Comité international des poids et mesures* (CIPM) and the General Assembly of the International Astronomical Union (IAU) defined Ephemeris Time as a uniform time scale, and redefined the second accordingly (Nelson et al. 2001; Sadler 1960). This definition was adopted when the International System of Units was launched by the 1961 *Conférence générale des poids et mesures* (CGPM). It was fundamentally astronomical: One second was *1/31,556,925.9747 of the tropical year for 1900 January 0 at 12 h ephemeris time.* This represented a fractional uncertainty of roughly $10^{-11}$ over an averaging time of $10^7$ s.

Atomic clocks, however, surpassed the level of precision attainable in astronomy (Fig. 1). As early as 1968, only thirteen years after Jack Parry and Louis Essen built the "Cæsium Mark 1" clock at UK's National Physical Laboratory (Essen and Parry 1955), the second was redefined in terms of atomic clocks as *9,192,631,770 periods of the radiation corresponding to the transition between the two hyperfine levels of the ground state of the cæsium 133 atom [at rest and at 0 K].*[1]

The technology of chronometric devices is improving steadily, and it is quite possible that a future CGPM will redefine the second in terms of optical clocks (Gill 2011).[2]

## *Uniform Time Scales*

A uniform time scale is defined in terms of a fixed unit of time. The latter may be realized with periodic phenomena or frequency standards. These can reach high accuracies but, generally speaking, the length of their continuous operation is limited.[3]

---

[1]The specification of conditions was added in 1997.

[2]It has been proposed that CGPM 2018 redefine all base units in terms of the second by fixing the numerical values of the constants *c, h, e, k,* and *NA* (Mills 2010). Although CGPM 2018 might redefine the second, it is more likely to defer the decision to the following CGPM.

[3]We shall refrain from discussing other vital components of every timekeeping system: record keeping, synchronization and time transfer.

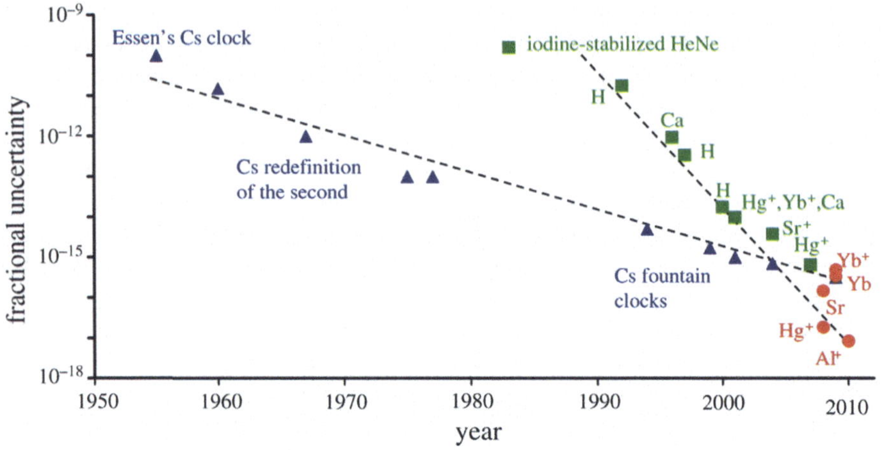


**Fig. 1** Progress of atomic clocks. Microwave clocks (triangles); optical frequency standards (squares); optical frequency standard with estimated uncertainties below Cs primary standard uncertainties (circles). Courtesy Gill (2011)

Atomic clocks allowed us to realize uniform time scales independently of astronomical observations. Early atomic clocks, however, had limited periods of continuous operation. On 13 September 1956 "A.1," the first atomic time scale, was launched at the US Naval Research Laboratory using a quartz-crystal clock the frequency of which was checked daily with an Atomichron cæsium standard.

The first uniform time scale, the Ephemeris Time, may be understood as the independent variable used in the calculation of the ephemerides.[4] Its realization, however, proceeds in the opposite sense: from astronomical observations to a uniform time scale. Thus, Ephemeris Time depends on the mathematical model of celestial mechanics applied in the calculations of the ephemerides. As a result, with more detailed models (e.g., including relativistic effects), the definition of Ephemeris Time became ambiguous. To allay these and other issues, several additional astronomical uniform time scales [UT1, UT2, TDT, TBD, TT, TCG, TCB; cf. McCarthy and Seidelmann 2009] were introduced.

In 1970, the *Comité consultatif pour la définition de la seconde* set forth guidelines for a new *Temps Atomique International* (TAI):

> International Atomic Time (TAI) is the time reference coordinate established by the Bureau International de l'Heure on the basis of readings of atomic clocks operating in various establishments in accordance with the definition of the second, the unit of time of the International System of Units (BIPM 1970b).

Following IAU recommendations (1967 General Assembly in Prague), TAI was set to agree with the astronomical time scale (UT2) on 1 January 1958 at 0 h UT2

[4]The apparent positions in the sky at given times of the Sun, Moon, the planets, and other celestial bodies, derived from our mathematical models of their motions.

(BIPM 1970a). The CGPM approved TAI in 1971 (BIPM 1972). To account for general relativity, the specification of the reference frame was added to the definition in 1980 (BIPM 1980).

## *Coordinated Universal Time*

The community was divided between those who required a broadcast time to reflect Earth's orientation in space, and those who preferred a perfectly uniform time scale. Coordinated Universal Time (UTC) was revised in 1972 to accommodate both demands.

UTC originally started on 1 January 1961 as the atomic time "coordinated" between US Naval Observatory, Royal Greenwich Observatory, US National Bureau of Standards, US Naval Research Laboratory, and UK's National Physics Laboratory, entrusted to the Parisian *Bureau International de l'Heure* (BIH) in 1965. Each of these institutions had also been "coordinating" their atomic time scale with Earth's rotation (as measured by UT2) by introducing step corrections. UTC continued this practice. BIH linked UTC to A3 (BIH's atomic time scale, which became TAI). Apart from step corrections, certain time scales had used *rate offsets* to "coordinate" with astronomy or with each other (they were not uniformly using the same unit of time). UTC was approved by IAU in 1967 (General Assembly in Prague; Nelson et al. 2001).

The present UTC was defined in January 1970 by the *Comité consultatif international pour la radio* (which became, after 1992, the Radiocommunication Bureau of the International Telecommunications Union), time and frequency recommendation TF.460-5 (Comité consultatif international pour la radio 1972), and was launched on 1 January 1972. Coordination with the International Astronomical Union was less then perfect, and the definition was tweaked after IAU's General Assembly of 1973.

UTC is a uniform time scale, founded upon the *Systéme international* base unit of time. Its offset from TAI is an integer number of seconds. Coordination with UT1 (the measure of Earth's orientation in space) is achieved with one-second step corrections, called "leap seconds" (at the end of June or December[5]), introduced by BIPM as instructed by the International Earth Rotation and Reference Systems Service (IERS) to maintain the difference $|UT1 - UTC|$ below 0.9 s (Fig. 2).

---

[5]In principle it could be at the end of any month but since 1972 it has always been only June and December, months which have been designated as first preference. March and September are the second preference.

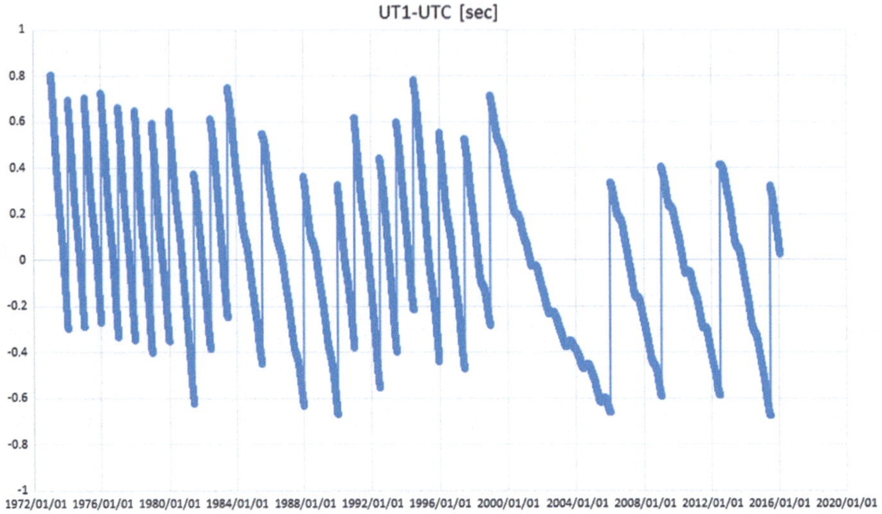


**Fig. 2** UT1–UTC. *Courtesy* IERS

## *The New Millennium*

In 1999, amid the millennial surge of interest in timekeeping, Dennis McCarthy and William Klepczynski (re)opened the discussion about UTC and its conformity with Earth's rotation (McCarthy and Klepczynski 1999). The primary concern of their paper had to do with global navigation satellite systems (GNSS; the oldest and best known is GPS operated by USA). When UTC was defined in the early 1970s, legal time was disseminated by radio. The ever-accelerating technological progress has brought about radical changes in this respect. Classic radio beeps have been replaced with the ubiquitous global navigation units and the Network Time Protocol (NTP, which keeps our internet-enabled devices synchronized).

New technology has brought about new issues. The "Y2K bug" was a major demonstration of the difficulties in ensuring that our software-dependent world survived rare conditions. In the light of how reliant we became on software, it was important to reevaluate priorities, and to consider minimizing irregular and unpredictable factors. Factors like the insertion of leap seconds into the otherwise uniform time scale appeared more problematic in this context than they did in 1970.

The same technology has introduced new possibilities. Distribution of time signals has become incomparably more efficient. Having several time scales available at one's fingertips has become commonplace.

Consequently, the need for a compromise time scale like the UTC presented itself in a very different light in this new context.

The paper (McCarthy and Klepczynski 1999) mentions another consideration. The authors examine the data on the evolution of Earth's spin rate, and suggest that they should be fitted with a parabola rather than a line. This would mean that the rate at which leap seconds will be needed might grow. They estimate that by 2050 we might need one leap second on average every 8 months. Whether this extrapolation is accurate or not, the argument is worthy of consideration. Leap seconds have been relatively infrequent up till now (26 leap seconds in 44 years) but there is no guarantee that they will remain so.

There is one argument, however, which is truly fundamental. It is that of metrological consistency. The community of metrologists institutionally values standardization. Many experts perceives it as their mission. After all, the universal adoption of the same units has been the worthy goal in this field since the French Enlightenment pursued Reason as the unifying force for the construction of the "brave new world', freed of sectarianism and all irrational divisions. I could put it no better than ITU's Working Party 7A in 2015.

> UTC without leap seconds will represent a continuous reference time-scale and will encourage the use of only one continuous reference time-scale making it truly universal. This will avoid the proliferation of other time-scales that may cause serious confusion, and contribute to the interoperability of systems (Working Party 2015).

McCarthy and Klepczynski listed several options:

1. Continue current procedure.
2. Discontinue leap seconds, fixing the difference between UTC and TAI at a certain integer number of seconds, and allowing the difference between UT1 and UTC to grow.
3. Change the tolerance for the difference between UT1 and UTC from the current 0.9 s to an unspecified larger value.
4. Redefine the second, linking the definition to the current or projected rotation period of the Earth but maintaining the general philosophy of the second as a constant interval.
5. Modify the current procedure by setting a time interval when inserting accumulated leap seconds would be allowable (McCarthy and Klepczynski 1999).

## *Debate 1999–2015*

We shall not go into the details of the debate which ensued. Let us limit ourselves to a cursory and incomplete list of formal meetings (excerpt from more comprehensive online resource[6]

In August 2000, the International Astronomical Union (IAU) created a Working Group to study the issue. IAU's Commission 31 (Time) was also involved.

[6]http://www.ucolick.org/~sla/leapsecs/ (Allen 2016).

In October 2000, official study of the proposal was undertaken by a *Special Rapporteur Group on the future of UTC* (SRG) established for the purpose by ITU-R[7] Working Party 7A.

In 2002, the SRG reported a "convergence of opinion" in favor of option 1 (above).

In May 2003, an ITU-R *Special Colloquium on the Future of UTC* took place in Torino. It was asked whether the a redefined UTC, non-compliant with the 1970 specification of $|UT1 - UTC| < 0.9$ s, should not be renamed to prevent confusion. No consensus was reached.

In June 2003, the World Radiocommunications Conference WRC-03 took place in Geneva. Further study was recommended.

In April 2004, the 16th meeting of CCTF[8] took place at Sévres. The SRG presented a "UTC Transition Plan". No consensus was reached regarding the leap-second mechanism and the name of the possible redefined time scale.

In September 2004, the ITU-R WP7A held a meeting in Geneva. SRG's "Transition Plan" notwithstanding, the representatives of United States of America unexpectedly proposed the leap hour as an alternative to the leap second. This would have been a concrete form of the option 2 above. Confusion ensued.

In July 2005, the IERS[9] in Paris solicited the response of those "affected by the content of the US [leap hour] proposal" using the same distribution mechanisms as IERS Bulletins C and D just after IERS Bulletin C, n. 30. This was followed by a similar move by UK's Royal Astronomical Society, which later also issued a press release (September).

In 2005, IAU's Working Group on UTC published its first report, stating that, "none of the options beyond [maintaining the status quo] has received significant acceptance."

In October–November 2007, WRC in Geneva reached no consensus, primarily because of UK's position. Further study was recommended.

In 2011, a colloquium on was held in Exton, Pennsylvania (Seago et al. 2011). The organizers invited new perspectives on the issue in order to examine all possible aspects of the issue and implications of the proposed solutions. In January 2012, ITU-R's Radiocommunication Assembly preparatory for WRC-12 asked Study Group 7 and its Working Party 7A to continue studying the future of UTC, inviting more national administrations to participate in the work of these expert bodies. (In real terms, WP7A had less than a dozen active members throughout its history.)

In May 2013, in continuation of the 2011 Exton colloquium, a meeting was held in Charlottesville, Virginia (Seago et al. 2013).

[7]ITU = International Telecommunications Union; ITU-R = Radiocommunications Sector of ITU.

[8]CIPM's *Comité consultatif du temps et des fréquence* (Consultative Committee for Time and Frequency). CIPM = *Comité internationale des poids et mesures* (International Committee for Weights and Measures).

[9]International Earth Rotation and Reference Systems Service.

In September 2013, a workshop was held in Geneva organized by ITU and BIPM.[10]

In January 2014, at the 223rd Meeting of the American Astronomical Society, there was a splinter session on the future of UTC.

In October 2014, the International Astronomical Union's *Working Group on UTC* first posted its report, and a few days later removed it from its web. A few weeks later it appeared on the IERS web. The positions of various IAU's bodies remained uncoordinated.

In August 2015, following IAU's 29th General Assembly in Honolulu, the Working Group on UTC and Commission 31 on Time were discontinued as a part of a major transformation of IAU's structure.

In November 2015, WRC-15 resolved not to change UTC at least until WRC-23 (Resolution COM5/1), and called for a broader participation of international organizations, especially the International Civil Aviation Organization, CGPM, IERS, the International Union of Geodesy and Geophysics, the International Union of Radio Science, the International Organization for Standardization, the World Meteorological Organization and the IAU. The underlying view seems to be that it is the BIPM who has the responsibility "for establishing and maintaining the second [...] and its dissemination through the reference time scale," while the ITU-R only "disseminates time signals via radiocommunication." (World Radiocommunication Conference 2015) The inference between the lines is that the (re)definition of UTC may be more appropriately in the purview of CGPM.

## Cosmic Cycles

The heavens are imbued with powerful symbolism. In most societies, they are identified with the eternal realm of gods and heroes, the dwelling place of the spirits of our ancestors. If we are to understand the leap second debate, we have to peer into the collective unconscious. This may be methodologically challenging in general but timekeeping is a singular case where I believe many salient points are easy to observe.

### *Purposes of Timekeeping: Practical and Otherwise*

Already around 380 BC, Plato mentions this evidence, "The observation of the seasons and of months and years is as essential to the general as it is to the farmer or sailor." (*Republic* 527c, Jowett 1892) Timekeeping (closely followed by navigation) is doubtless the most ancient mission of astronomy. It has an eminently

---

[10] *Bureau international des poids et mesures* (International Bureau of Weights and Measures).

practical purpose. This is all the more true today when, assisted by omnipresent accurate timepieces, our lives are meted out by the second under the dictate of schedules, agendas, timetables, milestones, and deadlines. Amid all our preoccupations we tend to forget that apart from these prosaic and utilitarian purposes, there is more to timekeeping than practicality, and however inane these other purposes may seem, they are much more profound and pervasive.

Time and eternity, each a separate enigma, are even more inscrutable when contemplated in their polar duality. As Plato's puts it, there is "that which always is and has no becoming," and there is "that which is always becoming and never is" (*Timaeus* 27e, Jowett 1892). On the one hand, our minds pine for the world of timeless truth, and on the other, we live in a world of relentless change, of birth, growth, decay, and death. The eternal and temporal are interlinked as are our minds and bodies.

As an example of how societies make the eternal reality present and potent in temporal activity, we shall evoke Mircea Eliade's classical description of the New Year rituals. Eliade identifies the fundamental motivation behind New Year's celebrations with a very archaic operation which he calls the Regeneration of Time (Eliade 1969). How ancient is it? It can be argued that it has been present for at least 60,000 years (Gabor 2011, 11–663).

## *Timekeeping Schemes: Artifacts with Symbolic Power*

Timekeeping schemes (calendars, UTC, the seven-day week…) are artifacts devised by individuals and societies. Regardless of their authors and their intentions, timekeeping schemes come to possess symbolic powers. Once they become symbols they gain a life of their own. The seven-day week, for instance, most likely started out in ancient Babylonia, subsequently became a part of the creation narrative of Genesis 1, and is now perceived as a symbol of Creation itself. Symbols have the ability to evoke mysteries and depths which hold a powerful subliminal sway over us. They draw us in and produce diverse emotional responses. *Even the institutions defining or modifying timekeeping schemes, have little control over the symbolic significance of what they regulate.*

## *Underlying Dynamics and Astronomical Conformity*

### Principle of Astronomical Conformity

Coupling timekeeping schemes with celestial phenomena has been, for the most part, a practical arrangement. Its justification, however, was overwhelmingly based on symbolism. St. John Paul II remarked, "we seek to blend the fundamental rhythms of human life in society with the fundamental rhythms of the universe in which we live" (St. John Paul II. 1982, XXII]. It can be shown on many examples from the history of timekeeping schemes (Gabor 2011, 11–663) that *although astronomical accuracy was often lacking in fact, what was universally present was the principle that feasts should be celebrated in conformity with natural rhythms in general, and astronomical phenomena in particular.*

Astronomical conformity was universally accepted as a principle but it was not always adhered to in fact. *The principle of astronomical conformity in timekeeping seems to fulfill its social function even when it is not perfectly observed in practice.* This discrepancy was sometimes caused by simple lack of astronomical knowledge (e.g., in the above-mentioned Chinese case), but more often by a conflict with other principles.

## Continuity and Timelessness

*Continuity* is a very practical aspect of any timekeeping scheme that often acquires a symbolic significance. A continuity since "time immemorial" confers upon a practice a certain air of timelessness, making it sacrosanct and thus immune to change. Social institutions enjoying an air of timelessness are often perceived as sacred rather than profane.

The Egyptian solar calendar was perceived as a divine gift and thus it was unthinkable to change it even though its non-conformity with the seasons soon became apparent (the year was 365 days long with no leap years). "The Kings of Egypt had to swear before they took office that they would not change the calendar…" (Richards 2000).

Whereas continuity is obviously important for practical purposes, it is by no means a sufficient reason for maintaining the highly impractical scheme, such as the continuous cycle of 7-day weeks. If practical considerations prevailed, a "weekless" day would have been inserted into the 365-day year long ago ($364 = 7 \times 52 + 1$). The continuous cycle of weeks resists change because, symbolically, it has been in place literally "since the beginning of time" (Second Vatican Council 1963). *What counts is the general perception, not the historical fact.*

## Inertia

Inertia is another property of social institutions which often hindered astronomical conformity. The overdue reform of the Julian calendar is a well-known example. The reasons why the change was so difficult to tackle ranged from simple procrastination to caution due to previous negative experience. *Inertia in such sensitive matters as timekeeping can be a force to be reckoned with.*

## Expediency: Shift from Empirical Timekeeping to Calculated Schemes

There is a general trend to replace empirical timekeeping with calculated schemes which do not follow the immediate astronomical phenomena but rather a uniformly progressing parameter corresponding to an average rate of motion over a certain period of time. A classic example is the shift from apparent local solar time to mean local solar time. This introduces a certain discrepancy between the Heavens and civil timekeeping as an unintended but inevitable consequence.

*What speaks in favor of the calculated schemes is expediency.* It is much easier to follow computational rules than to maintain an astronomical vigil over the celestial phenomena relevant for timekeeping. Empirical schemes need protocols and networks in order to announce the observed parameters, making it difficult to follow the empirical time. Calculated schemes can be extrapolated into the future, and into the past. The greatest disadvantages of the empirical systems is the need to keep detailed records in order to know how particular instances of time were labeled in the past, and the impossibility to predict the exact progression of the time scale in the future. *In the present case, debating the relative merits UTC and TAI is very much an instance of this general shift away from empirical timekeeping.*

By construction, *calculated schemes do not follow the principle of astronomical conformity strictly, rather, they respect it on average, minimizing long-term drifts.* Since computational rules cannot predict astronomical phenomena with perfect accuracy, *calculated timekeeping needs some correction mechanisms based on observation. Thus, at a certain level, calculated schemes also depend on observations.*

The major drawback is that such correction mechanisms *are rarely defined at the outset.* A positive exception is the Gregorian calendar. Gregory XIII's bull *Inter gravissimas* of 1582 envisages a tweak (without going into any specifics) when "the equinox recedes from March 21st," (Clavius 1603) and Clavius's *Explication* (Clavius 1603) estimates that this will have to be examined around the year 4000.

## Symbols and Reality

The relationship between symbols and "reality" is complex. Not so long ago, all human societies would have understood that the eternal realm of symbols and myths, values and virtues, Gods and heroes, *was the ultimate reality*, and our lives were a part of this *true* world only inasmuch we imitated these models, thus participating in what was perceived as timeless and therefore permanent (principally through ritualized behavior). Our society has consciously embraced a different notion of reality but our symbols are still anchored in the underlying ancient ontology.

*The collective unconscious perceives symbols as grounded in reality.* The symbolic function of timekeeping schemes is maintained work as long as they conform to astronomical phenomena at some level. The chthonic symbolism of *midnight*, for instance, has not been seriously unhinged by zonal time. Perceived astronomical conformity thus lends force to timekeeping symbolism. This relationship is mutual. *The symbolic value of timekeeping exerts a pressure on timekeeping schemes until they conform to astronomical Heavens.*

## Conclusion

Imagine somebody wants to do something you dislike instinctively. You cannot pinpoint the main reason why you hate it so much, you just know that you do. You know that you must do something to stop it. You are upset because of the proposal itself. You are agitated because you are compelled to think of countermeasures on the spur of the moment. You feel frustrated and rushed. And, adding an extra layer of unease, you are confused, suspecting that your reaction may not be fully justified because you cannot pinpoint why exactly you are reacting so strongly. The paradox is that consensus cannot be reached without a rational debate, and that can only happen once the parties get in touch with the whole issue, including the hitherto unnamed and unconscious concerns.

I find it particularly significant that the leap second debate has been inconclusive for so long. In fact, the only thing we know today is that it will not be over before 2023 (World Radiocommunication Conference 2015). I believe it is because most of the participants are in this enduring conversation are not quite in touch with all aspects of the issue.

I have been involved in some of the discussions of the international community of experts involved in the matter since 2011. My main purpose has been to facilitate the conversation by drawing attention to one dimension of the process in particular, viz., that of symbolism.

I believe that the proposal to abandon the leap second is likely to prevail because it is carried forward by the general trend described in Section "Expediency: Shift from Empirical Timekeeping to Calculated Schemes". I have tried to name the undercurrents involved, especially some of those that create turbulence (e.g., Sections "Principle of Astronomical Conformity, Continuity and Timelessness, and Inertia"). In order to moderate the subliminal reactions, I have suggested abandoning the naive rationalism of the Enlightenment's belief in the existence and attainability of incontrovertible definitive solutions (Section "The New Millennium").

In practice, this means refraining from saying anything that may sound like, "any rational person must see that this is the one definitive timescale." This statement illustrates two deplorable fallacies. First, people cannot be reduced to pure reason and it would be irrational to try. And second, there is no such thing as a "definitive timescale." The best we can do is to follow the example of Gregory XIII and Clavius (Section "Expediency: Shift from Empirical Timekeeping to Calculated Schemes") who introduced our calendar not as "definitive" but rather counted on future generations to improve it.

**Acknowledgements** This paper is the result of the author's participation in symposia held in Exton in 2011 (Seago et al. 2011), in Charlottesville in 2013 (Seago et al. 2013), and at Harvard in 2016 (Arias et al. 2017). It presents some material from (Gabor 2011) in a more popular form and with a different emphasis. The author is very grateful to the organizers of these meetings for their invitation and valuable discussions.

## References

Allen, S. (2016). *Future of leap seconds in UTC.* Online resource, last retrieved Nov 2016. http://www.ucolick.org/~sla/leapsecs/

Arias, F., Combrinck, L., Gabor, P., Hohenkerk, C., & Seidelmann, P. K. (Eds.). (2017). *The science of time 2016. Time in astronomy & society, past, present and future*. Springer.

BIPM. (1970a). Mise en pratique du Temps Atomique International. *Com. Cons. Déf. Seconde, 5*, 21–23.

BIPM. (1970b). Recommendation S2. *Proc.-Verb. Com. Int. Poids et Mesures, 38*, 110–111.

BIPM. (1972). Résolution 1. *Comptes Rendus de la 14e CGPM (1971)*, 77.

BIPM. (1980). Déclaration. *Com. Cons. Déf. Seconde, 9*, 15.

Clavius, C. (1603). *Romani Calendarii a Gregorio XIII. P. M. Restituti Explicatio. Clementis VIII. P. M. iussu edita.* Aloysius Zannettus, Roma.

Comité consultatif international pour la radio. (1972). TF.460-5 standard-frequency and time-signal emissions. Recommendations of the International Telecommunications Union, Radiocommunication Bureau.

Eliade, M. (1969). *Le mythe de l'éternel retour: archétypes et répétition*. Gallimard.

Essen, L., & Parry, J. V. L. (1955). An Atomic standard of frequency and time interval: A cæsium resonator. *Nature, 176*, 280–282.

Gabor, P. (2011). The heavens and timekeeping, symbolism and expediency. In J. H. Seago, R. L. Seaman, & S. L. Allen (Eds.), *Decoupling civil timekeeping from earth rotation*. American Astronautical Society.

Gill, P. (2011). When should we change the definition of the second? *Philosophical Transactions of the Royal Society of London A: Mathematical, Physical and Engineering Sciences, 369* (1953), 4109–4130.

Jowett, B. (Transl.) (1892). *The dialogues of Plato translated into English with analyses and introductions by B. Jowett, M.A. in Five Volumes* (3rd ed.). Oxford University Press.

McCarthy, D. D., & Seidelmann, P. K. (2009). *Time: From earth rotation to atomic physics*. Wiley.

McCarthy, D., & Klepczynski, W. (1999). GPS and leap seconds. Time to change? *GPS World*, 50–57.

Mills, I. (2010). *Draft Chapter 2 for SI Brochure, following redefinitions of the base units*. Paris: Bureau International des Poids et Mesures.

Nelson, R. A., McCarthy, D. D., Malys, S., Levine, J., Guinot, B., Fliegel, H. F., et al. (2001). The leap second: Its history and possible future. *Metrologia, 38*(6), 509.

Richards, E. G. (2000). *Mapping time: The calendar and its history* (p. 110). Oxford University Press.

Sadler, D. H. (Ed.). (1960). *Proceedings of the 10th General Assembly. Transactions of the International Astronomical Union.* CUP.

Seago, J. H., Seaman, R. L., & Allen, S. L. (Eds.). (2011). *Decoupling civil timekeeping from earth rotation*. American Astronautical Society.

Seago, J. H., Seaman, R. L., Kenneth Seidelmann, P., & Allen, S. L. (Eds.). (2013). *Requirements for UTC and civil timekeeping on earth*. American Astronautical Society.

Second Vatican Council. (1963). Declaration on the revision of the calendar. In *Documents of Vatican II*.

St. John Paul II. (1982). In G. V. Coyne, M. A. Hoskin, & O. Pedersen (Eds.), *Gregorian Reform of the Calendar. Proceedings of the Vatican Conference to Commemorate its 400th Anniversary, 1582–1982* (Address of august 31).

Tillich, P. (1964). *The nature of religious language* (pp. 53–67). Oxford University Press.

Working Party 7A. (2015). Agenda item 1.14. In *2nd Session of the Conference Preparatory Meeting for World Radiocommunications Conference 2015*, p. 57.

World Radiocommunication Conference. (2015). Press release, November 19, 2015.

# Practicing Science and Faith: A Short History of the Vatican Observatory

Ileana Chinnici

## The Origins

Historians agree in dating back the establishment of the Vatican Observatory to 1582, the year of issue of the document *Inter gravissimas*, announcing the reform of the Julian Calendar by Pope Gregorius XIII (1502–1585). This is due to the fact that, in order to test the calendar, some astronomical observations were carried out in the Vatican, in the Tower of Winds, where a meridian line was tracked on the floor by the famous cartographer Egnatio Danti (1536–1586).

However, the Tower of Winds cannot be considered an astronomical observatory in a strict sense, namely, a building with instruments and staff, conceived as a place where astronomical observations are carried out regularly.

In this sense, the first Vatican Observatory was the Collegio Romano Observatory, active from 17th to 19th century for regular research activities as well as for educational purposes. The Observatory was located firstly in the premises of the Collegio Romano (in Calandrelli's tower) and, later, under the directorship of Fr. Angelo Secchi (1818–1878), in new rooms erected on the roof of St. Ignatius' Church, where the renowned Jesuit astronomer carried out his early astrophysical studies.

Another astronomical observatory was established in 1827 by Leo XII (1760–1829) for the University of Rome; it was built on the tower of Campidoglio Palace and, from then, was in activity at the same time as the Collegio Romano Oservatory.

I. Chinnici (✉)
INAF-Osservatorio Astronomico di Palermo, Palermo, Italy
e-mail: chinnici@astropa.inaf.it

I. Chinnici
Vatican Observatory, Albano Laziale, Italy



G. Gionti, S.J. and J.-B. Kikwaya Eluo, S.J. (eds.), *The Vatican Observatory, Castel Gandolfo: 80th Anniversary Celebration*, Astrophysics and Space Science Proceedings 51, https://doi.org/10.1007/978-3-319-67205-2_16

However, after the annexation of Rome to the Kingdom of Italy, in the years 1870–1880 the properties of the religious orders were expropriated and the two Observatories were confiscated by the Italian Government.

## The New Vatican Observatory

Fr. Francesco Denza (1834–1894) (Fig. 1), barnabite, director of the astronomical and meteorological observatory of Carlo Alberto College in Moncalieri (Turin) did not resign to the idea that the Vatican State had lost its observatories.

Therefore, in 1888, he seized the opportunity of the jubilee of the priestly ordination of Leo XIII (1810–1903) to organize a Scientific Exhibition of the Italian Clergy. On that occasion, many scientific instruments were donated to the Pope

**Fig. 1** Portrait of Fr. Francesco Denza (from: Chinnici 1999)

and, at the end of the exhibition, Denza obtained the permission to install them in the Tower of Winds, the ancient Vatican astronomical site. That was the first step: once found a site and some instruments, he needed a sound scientific programme. This was provided by the *Carte du Ciel* project, an international enterprise promoted and headed by Paris Observatory, aimed at producing a photographic star catalogue and a chart fixing the image of the whole sky at the end of 19th century, in order to provide an extraordinary archive to the future astronomers. To plan the operations of this gigantic project—the catalogue should have contained all stars up to magnitude 11, while the chart stars up to magnitude 14—in 1887 an international congress was held in Paris: astronomical observatories of both hemispheres were invited to participate in the project and to use the same kind of instrument (the double—visual-photographic—Henry-Gautier astrograph). An international committee was also established in order to supervise and report the advancement of the work, with regular biennial meetings. A tentative list of participant observatories was prepared and the entire sky vault was divided into declination zones to be assigned to each observatory of the list.

In 1889, during the second astrographic congress, Denza come to Paris to negotiate the participation of the Vatican Observatory in the project. He wrote to Admiral Ernest B. Mouchez (1821–1892), director of Paris Observatory, a letter announcing (Fig. 2):

> … le Saint Père m'a chargé de la direction de l'Observatoire qu'on va organizer et batir dans le Vatican. Cet observatoire s'occupera presqu'exclusivement de la photographie celéste et surtout de la Carte du Ciel.
>
> (Denza to Mouchez: Paris, September 20th, 1889; in: Chinnici 1999, p. 461)

Not by chance, Denza appealed to France, a political allied of Vatican in those years, to take advantage of its support also on a scientific plan.

The participation in the *Carte du Ciel* was indeed also a political move: almost disappeared from the political scene after the annexation of its territories to Italy, the Vatican State intended to reaffirm its existence and identity by appearing in a board of nations sharing an advanced, daring, scientific project in order to reacquire prestige and visibility, at least in the scientific context.

France did not fail to support the Vatican participation: the Vatican Observatory was inserted in the list of the participants in the *Carte du Ciel* project and provided with a Henry-Gautier astrograph (Fig. 3), which was installed in the Leonine tower, inside the Vatican.

Consequently, in 1891 Denza attended the third astrographic congress in Paris, together with Fr. Giuseppe Lais (1845–1921) oratorian, his main assistant in the photographic work. The very same year, Pope Leo XIII formally established the new Observatory, the Specola Vaticana, by the decree *Ut mysticam*. It is interesting to read the text of the decree:

> So that they might display their disdain and hatred for the mystical Spouse of Christ […] those born of darkness are accustomed to calumniate her […] and they call her the friend of obscurantism, one who nurtures ignorance, an enemy of science and of progress […] everyone might see clearly that the Church and her Pastors are not opposed to true and solid

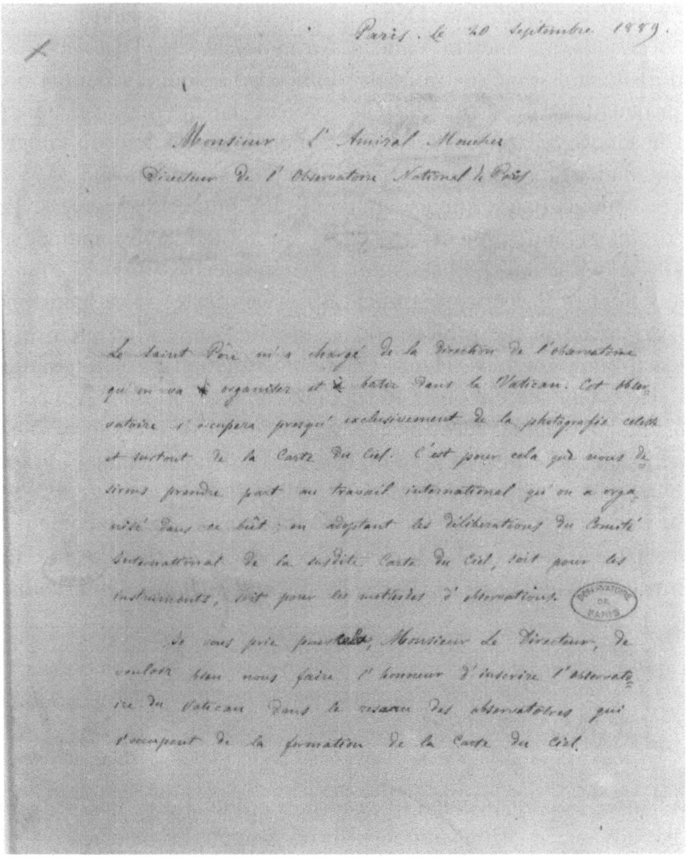
Paris, le 20 Septembre 1889.

Monsieur l'Amiral Mouchez
Directeur de l'Observatoire National de Paris

Le saint Père m'a chargé de la direction de l'observatoire qu'on va organiser et bâtir dans le Vatican. Cet Observatoire s'occupera presque exclusivement de la photographie céleste et surtout de la Carte du Ciel. C'est pour cela que nous désirons prendre part au travail international qu'on a organisé dans ce but en adoptant les délibérations du Comité International de la susdite Carte du Ciel, soit pour les instruments, soit pour les méthodes d'observations.

Je vous prie pour cela, Monsieur le Directeur, de vouloir bien nous faire l'honneur d'inscrire l'Observatoire du Vatican dans le réseau des observatoires qui s'occupent de la formation de la Carte du Ciel.



**Fig. 2** Letter by Denza to Admiral Mouchez (from: Chinnici 1999)

science [...] but that they embrace it, encourage it, and promote it with the fullest possible dedication.

(*Ut mysticam,* 1891; in: Maffeo 1991, pp. 207; 210)

From these words, it is clear that the Vatican Observatory was established with an apologetic intent—that of defending the Church from the accusations of obscurantism by anticlerical adversaries. The Church is here presented as a friend and protector of science, which is not opposed to faith, but offers another way to approach and contemplate the Creation. It is easy to recognize in the background the echoes of the anticlerical propaganda which violently attacked the Catholic Church in the second half of 19th century.

Denza died three years later, in 1894, but the work plan was carefully continued by Lais, who unceasingly worked at the *Carte du Ciel* project for about twenty years, almost accomplishing the photographic work related to the Catalogue (1040 plates, requiring an exposure of 5 min) and half of that related to the Chart (277 out

**Fig. 3** Fr. Lais (on the right) and the technician Federico Mannucci (on the left) at the Henry-Gautier astrograph of the Vatican Observatory (from: Chinnici 1999)

of 540 plates, requiring an exposure of 2 h) of the sky zone assigned to the Vatican Observatory.

In spite of some difficulties and delays—the photographic work for the chart plates was completed in 1953—the participation of the Vatican Observatory in the *Carte du Ciel* can be considered successful: it was among the few Observatories which completed the photographic work, the publication of catalogue and the printing of the chart plates. Many other observatories abandoned before starting, because of financial and technical problems, or left their work unaccomplished: therefore, the completion of all the assigned work was a remarkable achievement for the little but competitive Vatican Observatory.

## Consolidation and Expansion

After the death of Denza and the short directorship by Fr. Angelo Rodriguez Prada (1859–1935), an Augustinian meteorologist, who was removed because his competences in astronomy revealed to be insufficient for that position, the Vatican Observatory needed an expert and energic director, capable to achieve the imponent work related to the Carte du Ciel. In 1906 a new director was appointed, after the advice of reputed Italian and American astronomers: he was Fr. Georg Hagen (1847–1930) (Fig. 4), director of Georgetown Observatory in the US, who joined the Teutonic discipline and the American practical sense, together with a solid expertise in the field of variable stars' studies.

**Fig. 4** Portrait of Fr. Georg Hagen (from: Maffeo 1991)

Under his directorship the Vatican Observatory expanded and gained many more rooms and new instruments: at last four domes (and telescopes) were built on the Vatican walls (Fig. 5). Some newspapers, worried by this expansion, wrote that *the Jesuits invade the Vatican ...*

It is interesting to mention the reaction to this article by a renowned Italian astronomer, Michele Rajna (1854–1920), whose words confirm the high scientific reputation of Fr. Hagen among his colleagues:

> ... I usually have great respect for the opinion of others but that does not prevent me from protesting against the insinuation of that letter which is, in my opinion, a misguided attempt to carry anticlericalism even into an area which is so neutral, peaceful and objective, as is the study of the natural sciences in general and astronomy in particular, a field where those miserable human defects, which we call antipathy and prejudice, are – or at least should be – unknown at all. [...]
>
> The summons of Fr. Hagen in Rome was greeted by astronomers throughout the world with genuine unanimous applause [...]
>
> The scientific world will applaud the [...] denounced interference of the Jesuits in the affairs of the Vatican Observatory, when this interference consists in the forceful and wise direction of Fr. Hagen.
>
> (Rajna to the the Director of *Il Resto del Carlino*: March 7th, 1907; in: Maffeo 1991, pp. 47–49).

Fr. Hagen obtained from Pius X (1835–1914) also the Villa built for Leo XIII's summer retreats (today headquarters of the Vatican Radio) which was no more used: the offices of the Observatory were moved there, together with the prestigious

**Fig. 5** A general view of the Vatican walls with the domes of the Vatican telescopes: (1) Carte du Ciel astrograph; (2) Small Merz equatorial + small Merz refractor (lower floor); (3) Large Merz refractor; (4) Heliograph. On the right, (5) the old Tower of Winds (from: Maffeo 1991)

collection of minerals and meteorites donated by the marquis Adrien Charles de Mauroy (1848–1927) in the years 1896–1912.

In the years 1914–1928 Hagen completed the publication of the ten volumes of the Astrographic Catalogue. He paid less attention to the Chart: the photographic work had to be completed and the costs of the printing were very high but, above all, he probably was aware that their importance was declining, due to the spreading of new techniques, like the Schmidt cameras, having a wider photographic field. Moreover, in the same years, he was continuing his studies on variable stars, initially carried out at Georgetown College Observatory, and he was publishing the *Atlas Stellarum Variabilium* (1922–1941), later completed by his successor.

However, what is the most important, Hagen planned a new modern scientific programme for the Vatican Observatory: observations and studies of the Dark Clouds, then a major topic of the galactic astronomy. For this purpose, the support of Pius XI (1857–1939), elected Pope in 1922, was crucial.

## The Role of Pius XI

Pius XI strongly promoted science and technology: in 1931 he inaugurated the Vatican Radio and in 1936 he established the Pontifical Academy of Sciences, which is still today a very prestigious institution, gathering reputed scientists of any religious conviction.

**Fig. 6** Pius XI with Jesuit astronomers participants in the first IAU General Assembly; the third from the right is the Italian astronomer Vincenzo Cerulli (1859-1927), president of the Italian Astronomical Society, who joined the group (from: Maffeo 1991)

Since the beginning of his pontificate, the interest of Pius XI in astronomy was clear. The very same year of the election of Pius XI, the first General Assembly of the International Astronomical Union was held in Rome. The Union had been established two years earlier, in 1919, just after the end of the catastrophic World War I, and it was aimed at promoting cooperation among astronomers of all nations. Delegates from other Jesuit Observatories attended the Assembly and met at Vatican Observatory to welcome Pius XI, elected a few months earlier; the Pope visited the Specola and was portrayed in a photograph among the astronomers, in the hall of the large Merz refractor, as one of them (Fig. 6).

Consequently, when the increasing light pollution due to the urban expansion of Rome rendered impossible for the Specola astronomers to observe the dim glowing band of the Milky Way and risked to hamper the studies on Dark Clouds, the Pope was ready to offer his help.

## The Need of a New Site

The new Director, Fr. Johan Stein SJ (1871–1951), successor of Fr. Hagen, had to solve the problem of searching for a new site for the Specola. Pius XI initially supported the project of building a branch of Vatican Observatory in Ethiopia, where the seeing was considered excellent: in spring 1930 he sent a scientific expedition to test the site, but the expedition was stopped by the way because the political situation in Ethiopia had become unfavorable and the project was given up.

Moreover, it was necessary to renew the equipment of the Specola by acquiring modern telescopes, much more performing, to be placed in the new site. The scientific activity of the Specola risked to enter an impasse. Pius XI then made a surprising proposal. Stein wrote in 1932:

> The solution was found by His Holiness ... [who] would make available his Pontifical Villa at Castel Gandolfo [...] on the condition that the surroundings and the atmosphere should prove to be suitable for a branch of the Specola [...]
>
> Following the report of our experiments, it was in principle decided by H.H. that the branch of the Specola would be established and a new first-class photographic telescope acquired. [...]
>
> At the beginning His Holiness had called our attention to land in the huge gardens of the Villa Barberini, joined to Castelgandolfo. But quite soon his idea changed and he showed a preference to see the new telescope mounted on the top terrace of the Castle [...]
>
> [The] ideal circumstances and the spaciousness of the terrace led spontaneously to the idea of installing also the visual telescope there and so basically transferring the whole Specola to Castelgandolfo.
>
> (from: Maffeo 1991, pp. 79–80)

**Fig. 7** The domes of the Vatican Observatory at Castelgandolfo (from: Internet)

## Move to Gastelgandolfo

Moving the Specola to Castelgandolfo was a unique opportunity to establish a modern Observatory, thanks to the support of the Pope, who financed the purchase of new instruments. In the years 1932–1935, domes for a Zeiss equatorial (40-cm aperture, 600-cm focal length) and a Zeiss double astrograph (refractor 40-cm aperture, 240-cm focal length and parabolic mirror 60-cm diameter, 240-cm focal length Newtonian, 820-cm focal length Cassegrain) were built in the terrace of the Papal residence (Fig. 7).

The new Specola was also equipped with an important facility, a Spectrochemical Laboratory, with a large Steinheil spectrograph and many modern spectroscopic instruments. Fr. Alois Gatterer (1886–1953), who was in charge of the Laboratory, in 1933 described its scientific programme:

> The purpose of the laboratory is to contribute through quantitative and qualitative spectral research to the explanation of phenomena observed with the telescope and stellar spectrograph. It is not so much that we seek to know the composition of cosmic bodies but more importantly to know the physico-chemical conditions of the cosmic material [...] To accomplish this in the laboratory we try to modify the physico-chemical conditions of the luminous material so that its spectra will be comparable to the spectra observed for a heavenly object.
>
> In our research programmes we will especially focus on problems to do with the structure and composition of stellar and planetary atmospheres [...] Another focus of our research will be a spectral examination of the meteorites of the Specola [...] Furthermore, we will make an extensive investigation of the spectral emission of mixtures of incandescent gases which are of astrophysical importance.
>
> (from: Maffeo 1991, pp. 83–84)

In the first fifty years of activity, the Laboratory will produce many important scientific results, such as atlases of the spectral lines of iron, thorium, hafnium oxide and other molecular compounds. Gatterer also will promote in 1938 the publication of the international journal *Spectrochimica Acta*, whose third volume will be exceptionally edited by the Vatican, in the post-war years.

The inauguration of the new Specola took place on September 29th, 1935. Pius XI wanted that a stone was placed near the domes, with the inscription engraved: DEUM CREATOREM VENITE ADOREMUS (*Come, let us adore God the Creator!*). He pronounced an interesting talk, where his ideas about science and faith are well expressed. He clearly stated that the purpose of the Specola was to carry out astronomical research, in continuity with the tradition of his predecessors:

> … the acquired perfection of the scientific equipment and the proved scientific of men to whom it is assigned, will give […] some non trivial contribution to the study and the progress [of the science of the skies].
>
> […] the magnificent things which Astronomy studies […] are translated into matter of high spirituality …
>
> … what We do is not only to imitate and to continue […] the patronage incessantly praised of many Our distinguished predecessors […] it is also to resume … the thread of the many-centuries-old relationship [of the Roman Pontificate] with the science of the stars […] whose religious nature invites to prayer and worship.
>
> […] It seems to Ourselves … in this opening ceremony … that we make, on behalf of the whole Church, an act of Our priestly ministry.
>
> (from: Bertetto 1961, pp. 398–400)

The presence of an astronomical observatory in the Papal residence become an icon of the symbiosis between science and faith. Pius XI was certainly aware of the symbolic meaning of this choice and wanted the Specola hosted in his summer residence as a sign of the link between astronomy and spirituality, which he was convinced.

As it had been tested, the site of Castelgandolfo proved to be suitable for astronomical research and the Specola restarted the observations of the Milky Way structure with modern instruments and techniques.

To complete the *Carte du Ciel* work, Pius XI approved also the transfer of the Astrograph to Castelgandolfo; it was made in 1942, under the Pontificate of Pius XII (1876–1958): two domes were built in the gardens of Villa Barberini, the second one destined to a Schmidt telescope, built by Cox, Hargreaves & Thomson in London (spherical mirror diameter 98 cm, total length 500 cm), delivered in 1957. It is to be mentioned that, during the Second World War, the Papal gardens in Castelgandolfo hosted many refugees and that the staff of the Specola offered them a precious humanitarian aid. At last, the photographic work for the Chart was completed in 1953 and all plates were printed by 1955.

## The Last Fifty Years

The successors of Pius XI accepted the presence of the Specola in Castelgandolfo and often encouraged its activity. In 1965 a Computational Center was inaugurated at the Specola by Paul VI (1897–1978), but about a decade later, in 1976, the Spectrochemical Laboratory was shutted down because of the lack of staff. Again, the Specola needed a new equipment and a new site.

In 1980, the director Fr. George Coyne (b. 1933) signed an agreement between the Specola Vaticana and the Department of Astronomy of the University of Arizona, in Tucson, to use the Steward Observatory facilities and telescopes. This was the first step of a more ambitious project: the construction of a new telescope with modern technologies. In the years 1985–1991, the mirror for the Vatican Advanced Technology Telescope (VATT) was tested and constructed under the supervision of Fr. Chris Corbally (b. 1946) and its dome was built on Mount Graham, near the international Large Binocular Telescope (LBT). The construction of VATT was financed thanks to the generous support of a public subscription; this success led to the establishment of the Vatican Observatory Foundation, which still finances the telescope maintenance and service, as well as grants for young researchers and collaborators. The long wished extra-urban branch of the Specola was thus established overseas.

In 2009, International Year of Astronomy, the Specola has been unexpectedly moved from Castelgandolfo to Albano Laziale, in the premises of an ex-monastery of Basilian nuns, which has been entirely renewed on purpose. Astronomers now benefit from larger spaces for halls, offices, library, conference room, etc. while the telescopes remain installed in their original sites at Castelgandolfo Papal residence and are mostly used for educational and outreach purposes.

## Conclusions

Today the Specola continues his mission to embrace science and faith: the list of priorities for the decade 2015–2025 (I—Planetary Science, Astrobiology and Exoplanets; II—Stellar Evoluzion; III—Extragalactic Astronomy; IV—Cosmology) shows clearly that its scientific programme is aimed at exploring the frontiers of current astronomical research. The last director, Fr. José G. Funes (b. 1963), summarized the spirit of the work of the astronomers at Specola:

> We are enthusiastic about our mission. Like all astronomers, our deepest desire is to be on the frontier of astronomical research; we share with our colleagues the same excitement in seeking answers to the fundamental questions about the universe: are we alone? Are there other Earths? How do stars and planets form and evolve? How galaxies form and evolve? What is dark matter and dark energy? What do we know about the universe in its first instants? Are there many universes? Sometimes I have been asked if the Vatican officials

> give us an agenda for our research goals. Our only goal is to do good science; our only commitment is to pursue truth, wherever it is to be found.
>
> (Funes 2014, p. 2).

The recent appointment of br. Guy Consolmagno (b. 1952), renowned expert in the field of meteorites' studies, as director of the Specola, reinforces the link with overseas and confirms the intent of *doing good science* which marks the history of the Specola.

## References

Bertetto, D. (1961). *Discorsi di Pio XI* (Vol. 3, 1934–1939). SEI.

Chinnici, I. (1999). *La Carte du Ciel. Correspondance inédite conservée dans les Archives de l'Observatoire de Paris*. Observatoire de Paris & Osservatorio Astronomico di Palermo "G. S. Vaiana".

Funes, J. G. (2014). Preface. In *Science Priorities of the Vatican Observatory for the next decade*.

Maffeo, S. (1991). *In the service of nine Popes: 100 years of the Vatican Observatory*. Vatican City: Libreria Editrice Vaticana.

# Part VI
# Science-Philosophy-Theology

# Some Reflections on the Influence and Role of Scientific Thought in the Context of the New Evangelization

**G. Tanzella-Nitti**

**Abstract** In this paper, after discussing how scientific culture shapes the way of thinking of great part of contemporary society, I briefly investigate the relationship between science and non-believing, showing that the latter cannot be presented as a direct consequence of the former. Rather, this is an ideologically-clothed popularization of science, which presents scientific culture as opposed to Christian faith. In order to evangelize a society that is highly shaped by the rationality of science, when addressing the relationship between faith/theology and science, a number of *clichés* must be overcome. At the same time, however, specific and positive aspects of scientific culture, mainly the humanistic and spiritual dimensions associated with research work, have to be highlighted. In order to foster a New Evangelization in the world of science, I suggest developing five leading ideas on the nature of scientific activity as such, and I single out four proposed tasks for scientists that are also believers and for pastors and theologians.

## Science, Secularization and New Evangelization

Countering the general disengagement trend of postmodernist thought, scientists are seen by the Catholic Church as learned interlocutors, who embody specific rational needs, despite being rightly or wrongly associated with agnostic or atheistic ideas. I believe that within the 'New Evangelization' task set by the Catholic Church at the beginning of the third millennium, the encounter with scientific culture is to be seen not only as a challenge, but also, and even more, as a significant *opportunity*.

---

To be published in the Proceedings of the Vatican Observatory 80th Anniversary Symposium, Springer 2017.

---

G. Tanzella-Nitti (✉)
Faculty of Theology and SISRI—Scuola Internazionale Superiore per la Ricerca Interdisciplinare, Pontificia Università della Santa Croce, Rome, Italy
e-mail: tanzella@pusc.it



G. Gionti, S.J. and J.-B. Kikwaya Eluo, S.J. (eds.), *The Vatican Observatory, Castel Gandolfo: 80th Anniversary Celebration*, Astrophysics and Space Science Proceedings 51, https://doi.org/10.1007/978-3-319-67205-2_17

There are some knots which have to be untied. Firstly, we have to make it clear that scientific culture cannot be hastily associated with atheistic or agnostic thought. Such an association, which is amplified by the media, has an ideological origin. This can be shown both within a theoretical framework, by examining the relationship between faith and reason historically forged by Christianity, and within a phenomenological-existential context, by turning to the history of scientific thought and to several scientists' biographical profiles.

The second critical knot concerns the relationship between scientific culture, technological progress and secularization, taking the latter to be a positive hurdle to the spread of the Gospel and to be synonymous with materialism. As the theologian Jean Danielou put it years ago: "If secularization is tantamount to the gradual disappearance of a more or less mythical view of the universe, where scientific advances teach us to make a distinction between primary and secondary causes, then I would say that in this case secularization is a merit of modern culture. In that sense, it is by all means clear that none of us can deny that secularization is a wonderful achievement. It would be absurd to even attempt to oppose science simply because it replaces some mythical representations or certain magic rituals. If, though, secularization is interpreted as meaning that from now on the scientific way of accessing knowledge would become the only kind of knowledge, thereby implying the end of metaphysics and the beginning of the dictatorship of the sciences, then I would say that this would amount to a frightening cultural regression. The universe may well be the object of scientific knowledge and at the same time continue to be the starting point of metaphysical knowledge; in other words, it may well lead us to know something different from its own sheer phenomenological laws."[1] As a commonplace of a number of theological analyses, the inevitable association of science with secularization is all too often taken for granted on the basis of a view of science and technology which, since Max Weber and, later, Martin Heidegger, has influenced sociological and philosophical thought, engendering a view that in Jean-François Lyotard has by now turned into a set phrase: "I define postmodern as incredulity toward metanarratives. This incredulity is undoubtedly a product of progress in the sciences: but that progress in turn presupposes it."[2]

Typically, to back up the above argument, reference is made to some influential figures of 18th-century Enlightenment and of 19th-century materialism. Considering scientific progress a cause of disbelief and a driver of secularization has thus become a view that is easily supported by most authors: initially referred to as *deviations* from a scientific mind-frame, secularization and materialism end up tragically turning into synonyms of 'scientific mind-frame,' or sometimes even of 'scientific method.' This generates a way of thinking that in many cases gets to

[1]Danielou (1972).

[2]J.-F. Lyotard, *The postmodern condition: a report on knowledge,* translation from the French by G. Bennington and B. Massumi; foreword by F. Jameson (Manchester: Manchester University Press, 1994), xxiv. Original French edition: *La condition postmoderne* (1979), Italian translation: La *condizione postmoderna. Rapporto sul sapere* (Milano: Feltrinelli, 1991), 6.

influence the theoretical and cultural contexts within which Pastors are to discuss the relationship between science, society and the Church.

Such a state of affairs is particularly harmful both for theology and for Christian faith, being unable to intelligently consider (*intus legere*) science's endeavor, which gets pushed to one side as a liability and is not seen, as it should, as an asset in the balance sheet of the field forces working for a new evangelization. The very position of the Magisterium of the Catholic vis-à-vis science is perceived by public opinion above all in terms of cautious watchfulness vis-à-vis technological, more specifically bio-technological, applications. This position is usually associated to a renewed self-criticism regarding certain historical events of the past, which would periodically be remembered to state that the circumstances which resulted in those mistakes are no longer applicable. Although the reflections of the Magisterium of the Church on sciences and technological-scientific activities are more substantial and complex, in general public opinion moves along different lines, almost exclusively underlining defensive attitudes or those favoring a peaceful and definite separation.

A reflection on the advancement of a new evangelization effort targeted at scientific culture in turn calls for a reflection on the roots of the predominant negative view of science. Is it only a matter of communication strategies, or is the relationship between Christian faith and scientific thought mediated, and sometimes filtered, by prejudices affecting their mutual dialogue not only within public debate but also in the context of theology and of the Church? Scientific work and its outcomes, we should bear in mind, are popularized by the media, swinging the image of science between triumphalism and catastrophism, presenting it either as a solution for all the problems of humankind or as a cause of imminent self-destruction. This kind of mediations and prejudices may have negative impacts on an underequipped theology and on deficient religious teaching, generating uncertainties which sometimes creep into ecclesial reflection or into pastoral planning documents. Even though, on the one hand, an in-depth knowledge of a culture is always the first step towards the inculturation of faith into new peoples and contexts, on the other hand, the evangelization of scientific culture cannot ignore regaining a sufficient familiarity with the language and the contents of sciences. The Catholic faithful active in the scientific environment certainly are familiar with such a language and contents, but pastors and theologians tasked with orienteering and serving its action do not seem to be.

## The Influence of Scientific Culture on the Proclamation of the Gospel

When discussing evangelization and inculturation of faith in relation to the technical-scientific context, we are not referring to a cultural *élite* or to a niche of experts to whom we talk about God resorting to intellectual parameters which may not even interest the majority of the community. Rather, in keeping with some well-known reference points in the Second Vatican Council documents,[3] the task in question involves wide strata of contemporary society to whom the twenty-first century Catholic Church wishes to restart proclaiming the mystery of Jesus Christ, crucified and risen, as the center of the universe and of history[4]—universe and history as visible to anyone and nowadays mostly judged through scientific categories, ones that evangelization cannot overlook. So, let us look at some aspects of the influence that scientific culture exerts on Gospel propagation.

First of all, in the countries of the globalized world scientific knowledge nowadays represents the implicit cultural context for any parties involved in a dialogue. Whether they are scientists or not, the recipients of Gospel's proclamation have got a *forma mentis* [a 'frame of mind']—this is the phrase used by *Gaudium et spes*, n. 5—that is significantly shaped by the achievements of the sciences. The latter are held to be an authoritative source of knowledge, often the most authoritative one; scientists and researchers are well-received by the general public even when they talk about social and moral issues. Scientific thought provides an ever higher number of people with a reference framework to evaluate statements, situations and events. On social networks it is not rare to see that the most widespread non-believing opinion is precisely that religion—particularly Christian religion—would no longer hold, especially when confronted with the new knowledge afforded by the sciences.

Secondly, there are several achievements of contemporary scientific research which call on Christian theology to elaborate new and sound syntheses between faith and reason. Nowadays, a number of teachings drawn from biblical Revelation need to be presented through a compelling hermeneutics suited for those who are familiar with the context of the natural sciences, of psychology and history. This calls for an in-depth analysis which theology or catechesis did not require right up to a few decades ago. Think, for instance, of the timeframe going from the appearance of *Homo sapiens* on the earth right down to the rise of the earliest oral traditions collected in the biblical narratives of the origins, including those concerning the primeval revelation and the original moral fall. Think of the morphogenetic and phylogenetic place of human beings within the extended evolution of life on our planet, also in relation to what caused that evolution; think of the possibility of providing a scientific description of many aspects traditionally associated with a human being's spiritual life, such as emotions, feelings and the neuro-physiological dimensions of free will; think also of the huge space-time cosmic scenarios in which we now know our tiny planet to be located, forcing us to change radically the categories of human history, up to make it plausible for life (and intelligent life as well) to be present in contexts other than the planet Earth. And also think of the questions posed by Christian eschatology concerning the link between history of the universe and history of salvation. Finally, in the longer term,

[3]Cf. Vatican Council II, *Gaudium et spes*, nn. 5, 33.

[4]Cf. John Paul II, *Redemptor hominis*, March 4, 1979, n. 1.

one ought to consider the possibility—no longer that remote—of synthetizing living organisms in a laboratory, along with the trans-humanist thinkers' push to act on the evolution of the human species, with changes breaking with the past and opening up completely unprecedented scenarios. For many of these questions theological and philosophical thinking can take—and is actually already taking—some viable paths; however, these scenarios must not be set aside as futuristic fruits of the imagination, simply because theologians do not know their language or implications. The time and ways these issues are to find their place in the theologian's schedule will depend on many factors, but clearly, sooner or later, given the scientific culture we are now moving in, they will inevitably have to be tackled.

A third aspect of how science culture impinges on evangelization consists in the way scientific applications have changed and continue to change the life of individuals and society. It is apparent to all that the relationships between human beings, but also the labor world and market, the education of the new generations and our relationship with things have deeply changed as a result of the information technology revolution, of he wide-spread virtual reality operating context, and of the new opportunities offered by global communication. The newly arising context will also necessarily impact on the way human beings understand themselves and the meaning of their relationships with others, affecting their intellectual, emotional and relational spheres. The applications of sciences that are changing our way of living include the new biomedical and biotechnological applications, as well as robotics, domotics and the gradual integration between human functions and operations entrusted to machines. Even though the way of looking at these new contexts aims to highlight the underlying ethical issues, one should not forget that this transformation first entails a new relationship between the human beings, their potentialities and expectations. The Gospel message, therefore, has to deal with what man can expect of technology, what could he trust and entrust to it—all aspects now closely related to human happiness and aspirations, as well as to how to live and die.

## Views of Science Around the 2012 Synod on the New Evangelization and Pope Francis 2013 Document *Evangelii Gaudium*

In order to better evaluate the role scientific culture may play in the context of a New Evangelization, it is instructive to look back at what happened a few years ago on the occasion of the 2012 Synod of Catholic Bishops devoted to the New Evangelization for the transmission of the Christian faith. Actually, technical-scientific research was one of the six sectors mentioned by that Synod and proposed for reflection to the general Assembly. In a document issued to help Bishops prepare their talks we read the following statement: "The fifth sector is scientific and technological research. We are living at a moment when people still marvel at the wonders resulting from continual advances in scientific and technological research. All of us experience the benefits of this progress in our daily lives, benefits on which we are becoming increasingly dependent. As a result, science and technology are in danger of becoming today's new idols. In a digitalized and globalized world, science can easily be considered a new religion, to which we turn with questions concerning truth and meaning, even though we know that the responses provided are only partial and not totally satisfying. New forms of 'gnosis' are emerging where technology itself becomes a kind of philosophy in which knowledge and meaning are derived from an unreal structuring of life. These new cults, increasing each day, ultimately end up by turning religious practice into a clinical form of seeking prosperity and instant gratification."[5]

The scenario outlined here mainly refers to the influence that scientific thought exerts on contemporary society's way of thinking and living, especially because of the images of science portrayed by the mass media and in public debate. This preliminary document expressed the concern that science could be raised to a new religion by the imposition of its method of acquiring knowledge on other areas of reality and by the allure of materialistic models due to the excessive trust widely put in technical capability.

Speaking of science pointing out the deviations of scientism or warning about the dangers arising from technology neglecting the good of man, ultimately expresses understandable concerns. However, in my opinion, this could end up by endorsing a view of scientific activities tending to set a dialectic opposition between science and ethics, science and wisdom, science and religion. Christian faith would then be tasked with reminding science about its limitations, its shortcomings, and its ever-present risk of rising to a criterion of interpretation and judgment of the whole of reality. Underlying such concerns is undoubtedly a legitimate point of view. Yet, if this perspective were not completed by looking at science from other angles, that view would not integrate all dimensions of scientific activity, for instance as they were highlighted by both the Second Vatican Council, and more recently by the teachings of John Paul II and of Benedict XVI, the latter being less extensive but equally profound. Scientific achievement—as the Magisterium of the Church has repeatedly stated—is indeed an achievement of truth, positively contributing to learning about the cosmos and man's own role within it, with doubtless potential for serving humankind and the quality of its life on earth. Science possesses significant humanistic dimensions which qualify it as a value in itself, namely a spiritual value.

In fact, as early as the middle of the 20th century, top scholars have demonstrated how science remains 'open' to its philosophical foundations; science is not a closed, self-referential kind of knowledge—and, in this sense, true science cannot become an ideology. It presupposes not only logics, but also ontology and a philosophy of nature. Moreover, authors such as Wittgenstein, Gödel, Tarski or Turing have demonstrated that the need for logical and ontological foundations is perceived 'from within' the formal language of science. Scientific reason 'extends'

[5]Secretary of The Synod for the New Evangelization, 2012, *Lineamenta*, n. 6.

itself to the point of requiring the introduction of typically philosophical notions within the horizons of science. These new openings prove some degree of convergence with what the magisterium of Benedict XVI has repeatedly emphasized regarding the urgency of a 'widening reason.' In this sense the sciences seem to offer noteworthy opportunities, thanks to their rigorous and demonstrative nature. This is an occasion, a sign of the times, which should not be overlooked.

After the work of the 2012 Synod, on November 24, 2013, Pope Francis issued a document entitled *Evangelii gaudium*. Although the Holy Father refers to some of the propositions approved in that Synod, he develops his own thought in quite a personal fashion. One of the passages of most interest for science is perhaps the following: "Proclaiming the Gospel message to different cultures also involves proclaiming it to professional, scientific and academic circles. This means an encounter between faith, reason and the sciences with a view to developing new approaches and arguments on the issue of credibility, a creative apologetics, which would encourage greater openness to the Gospel on the part of all. When certain categories of reason and the sciences are taken up into the proclamation of the message, these categories then become tools of evangelization; water is changed into wine. Whatever is taken up is not just redeemed, but becomes an instrument of the Spirit for enlightening and renewing the world."[6]

According to Pope Francis' remark, scientific categories and insights can become (I would say must become) a tool for evangelizing, thereby turning the water of science into the wine of truth, the only and one truth to which faith and reason together belong. In this paper I would offer some suggestions along these lines, reflecting upon what positive role scientific culture could play in a new evangelization. I wish to underline that science cannot be considered as a source of trouble for faith or for the Church, but rather as an ally and a fascinating partner. In a word, scientific culture is a sector of the present century's life offering the Church important opportunities.

## Some Suggestions for Scientists and Theologians

When we speak of a New Evangelization, we should remember that the subject entrusted to evangelize scientific culture is not only the Church through her pastoral documents or the events she organizes in a somewhat institutional way. The subject able to proclaim the Gospel to a scientific environment is any Christian faithful who acts and works in the scientific world. Contrary to what the mass media perceive and spread, the number of believers, including Catholics, working in the field of scientific research is meaningful. However, believing scientists need the support of pastors and the necessary assistance of philosophers and theologians to reach a synthesis between faith and reason, a synthesis that often remains difficult to

[6]Pope Francis, *Evangelii gaudium*, November 24, 2013, n. 132.

achieve to them, especially due to the lack of a philosophical and theological training.

In order to overcome a neutral and instrumental view of science and to promote a view of scientific culture capable of joining in the Church's task for a new evangelization, I suggest taking into account and developing the ideas listed below.

In the first place, we should remember that scientific enterprise participates in the human journey toward the truth. Albeit within the boundaries proper to its own method, science perceives some light of the presence of the Logos, by whom and through whom all things were made. When properly presented, the emphasis on a bond linking scientific enterprise with truth operates an important countertrend to relativism and indifferentism: nature is worthy of being studied, it is a source of beauty and meaning; it exerts an appeal towards the search for truth.

Scientific enterprise reveals and increases human dignity; science is not an impersonal and purely objective activity: it is a value in itself. Science is a body of knowledge worthy of being taught and transmitted, an important source of education and training also in spiritual values. Man has an undeniable vocation to the unity of knowledge; science plays an important role in achieving such a unity. Without a convincing synthesis between reason and faith, between what I study on nature and what I believe about nature, the evangelization of learned men and scholars would be impossible, or would only remain superficial.

In the second place, we should emphasize that science has a tremendous capacity for the achievement of the common good and for the development of peoples. Scientists, precisely because they know more, should serve more. When talking of science or technology, we should not just remember the risks of biology, biotechnology, or nuclear physics, but we also have to mention the enormous potential science and technology have for the common good, thus balancing sometimes catastrophic visions transmitted by the media or present in public debate.

It is also important, in my opinion, to overcome some common *clichés*, often used to express the relationship between the 'reading' of material reality provided by science, and the 'reading' of the created world provided by Christian revelation. Too strong a separation between these two 'readings,' the scientific and theological ones (NOMA, that is, Non Overlapping MAgisteria perspective), could be misleading, because it would confine the latter within the context of myth, or would only see the theological content of the Bible as a totally subjective account. This attitude could end up enhancing an already widespread fideism, especially among scientific researchers who are also believing scientists. At the same time, naïf concordism or attempts to seek a foundation or a 'demonstration' of faith in science should be avoided too. However, while *concordism* is certainly to be avoided, the existence of *consonance* and harmony between faith and science perspectives must also be affirmed.

When promoting the evangelization of scientific culture, it is important to remember the positive historical role played by Christian theology, creation theology in particular, for the birth and development of the scientific method in the Western culture. The unreliability of those judgments charging the Church with having hampered the development of science must be defended and clarified. In relation to that, some questions associated with the Galileo affair and the legacy of Giordano Bruno must be properly explained. These are two issues to which men of science are particularly sensitive and which are often exploited for ideological purposes, causing a major obstacle to the spread of the Gospel; and this is also because of some historical and epistemological ignorance shared by the general public. Today, the position of the Catholic Church regarding the beginning of human life and her denial to manipulate human embryos must also be thoroughly explained and grounded, since the general public erroneously considers as scientific activity what actually belongs to market or economic strategies.

Finally, here are some suggestions for scientists who are also believers, for pastors and theologians. I think that the New Evangelization in the context of scientific culture may largely depend on how we succeed in putting them into practice.

In proclaiming the Gospel, the example of people who were sincere men and women of faith and good scientists must often be mentioned and highlighted. There are many suggestive historical examples in this respect. There is no shortage of testimonies, but they must be made known to faithful Catholics and to the public at large.

Catholic scientists should not limit themselves to ‘being present’ in the world of science, but they are also called on to ‘evangelize scientific research’ from within, steering it towards truth and goodness. To this end, Catholic scientists are encouraged to sincerely seek the unity of knowledge, by gaining greater insights into those aspects of their faith that have a major relationship with their scientific research, thus achieving a higher synthesis between faith and reason. The first and most important evangelizers in the technical-scientific environment are not pastors, nor theologians, but the lay faithful that are professionally active in scientific research and in those places where this culture is forged.

On their side, Pastors must prepare themselves to proclaim the Gospel in a contemporary society which is highly influenced by the rationality of science. It is hoped that in the future the institutional studies training pastors for the priesthood and beyond will pay greater attention to scientific results and to scientific thought in general. This is especially necessary in those geographic areas particularly involved in the task of a new evangelization, where scientific culture has become part of the way of thinking and judging of a very broad range of people.

Finally, theologians’ interest for science is very welcomed. In dialoging with science, theologians are not only invited to study the compatibility between scientific results and biblical Revelation, but also to make use of proven scientific knowledge as an aid to better understand the Word of God. In this way the proclamation of the Word will become more profound and meditated and, therefore, more effective and helpful.

The above-listed suggestions are certainly demanding. Yet, they are all mentioned and contained, in a seminal fashion, in the exhortations of the Second Vatican Council, and they have all been personally exemplified by qualified actors all along the history of theology and of the Church. The value of scientific enterprise and the role it plays in the progress of humankind, in the search for truth and for the good, cannot be underestimated. Aware of that Pastors and theologians are called to help scientists, believers or non-believers, to discover the dignity of the role they play in society and in the Church. This was well grasped by one of the sharpest commentators of *Gaudium et spes*, Henri de Lubac, who exhorted modern apostles as follows: "They should not be led, through fear of the consequences drawn by atheism, to depreciate science or to curse technology; rather, they should not link them with a denial of faith they do not entail at all. Let them prove true friends of those who may have been misguided on the way; let them, then, offer them a hand and invite them to continue along the way together, till the end, when new light will shine on both of them."[7]

## References

Danielou, J. (1972). La secolarizzazione e la non credenza. In Caporale & Grumelli (Ed.), *Religione e ateismo nelle società secolarizzate* (p. 76). Bologna: Il Mulino.

De Lubac, H. (1985). Confronto con l'ateismo contemporaneo. In *La rivelazione divina e il senso dell'uomo* (p. 223), Milano: Jaca Book.

[7]De Lubac (1985).

# The Philosophy of Expertise: The Case of Vatican Astronomers

**Louis Caruana, S.J.**

The philosophy of expertise represents a new interesting research program in the area of science and religion. It is particularly significant because it refers not to particular scientific discoveries, considered individually, but to the general dynamics of the practical side of science. The crucial concept, expertise, involves the appeal to authority for the justification of arguments. Many people like to think that scientific knowledge is far from all appeal to authority. Those who are directly involved in scientific practice however know that science is not so clean. Appeal to authority remains very much part of the natural sciences, whether we like it or not. Some philosophers have recently ventured into these somewhat dark caverns of scientific thought and practice, and the result of their work has become very significant. In this chapter, I will first offer a brief overview of this philosophical work, and then explore how the new insights regarding scientific expertise and scientific appeal to authority can throw light not only on how science works but also on the issue of authoritative knowhow within the Church. To avoid getting lost within the world of abstract principles, I will focus on one particular concrete case, the case of Vatican astronomers. This is a particularly interesting case because this group's complex role lies precisely at the intersection between theological and scientific expertise. We can focus on one concrete question that allows us to see how such abstract principles work. Our target question is "For policies involving scientific issues, can ecclesiastical decision makers like Bishops in Rome ever be genuine experts and genuinely authoritative when they are not scientists?" It should be evident that I intend this question to represent difficult episodes such as the 1968 encyclical *Humane Vitae* in which Paul VI rejected most forms of birth control, the 1983 US Catholic Bishops' Pastoral Letter on War, Peace and Nuclear deterrence, and some of the current mounting criticism against Pope Francis's *Laudato Si*. In

L. Caruana, S.J. (✉)
Faculty of Philosophy, Pontifical Gregorian University, Piazza della Pilotta 4, 00187 Rome, Italy
e-mail: caruana@unigre.it



G. Gionti, S.J. and J.-B. Kikwaya Eluo, S.J. (eds.), *The Vatican Observatory, Castel Gandolfo: 80th Anniversary Celebration*, Astrophysics and Space Science Proceedings 51, https://doi.org/10.1007/978-3-319-67205-2_18

these and similar cases, people criticize the Church for allegedly sticking its nose into scientific issues where it has no competence.

## Expertise in Scientific Practice

Current work on the nature of expertise within science is the fruit of a long philosophical project that emerged from the work of Ludwig Wittgenstein, especially his insights about the importance of what people do for determining what they mean. Practice is the bedrock of semantics. Meaning is use. Such inspiring ideas were applied to various philosophical areas of inquiry by Wittgenstein's successors, one of whom, Gilbert Ryle in his 1949 book *The concept of mind*, derived the important epistemological distinction between *knowing-that* and *knowing-how*. The former kind of knowledge is propositional. We can factor out *knowledge-that* piecemeal into bits of information, with various kinds of inferential links between them. This is our normal view of knowledge. The other kind of knowledge, *knowledge-how*, is involved in the acquisition of practical skills: knowing how to the sail a boat, knowing how to ride a bike. Ryle's bold affirmation, one that we now recognize as a breakthrough, was that *knowledge-how* cannot be reduced to *knowledge-that*, in spite of our impression that it can. If you know how to ride a bike, it is not because you have learnt a long set of propositions or rules to guide you in all foreseeable situations. We say you acquire a skill precisely because what you acquire lies at a different level. If you try to think through your cycling, chances are that the propositions would confuse you and you would topple over.[1] From here, we can see the possibility of having some individuals becoming very good in a skill, and thereby becoming a model for others. We call them experts. The last phase in this very quick historical sketch I am offering came with the pioneering work of Michael Polanyi whose books *Personal Knowledge* (1958) and The *Tacit Dimension* (1966) brought Ryle's discussion to the heart of scientific practice. These works remain very relevant today, especially because of their defence of the idea that scientific work, even though guided by rules, relies on personal judgements. Science is not an engine that churns out truths blindly. Even its most formalized areas depend on prior personal commitments and social factors.[2] We believe more than what we can prove, and we know more than

[1]For the current state of the debate on the nature of knowing-how, see J. Bengson and M. A. Moffett (eds.), *Knowing How: Essays on Knowledge, Mind, and Action*, Oxford University Press, 2012; David Lewis, "What Experience Teaches," in *Mind and Cognition: A Reader*, W. G. Lycan (ed.), Oxford: Blackwell, 1990, pp. 469–477; Paul Snowdon, "Knowing How and Knowing That: A Distinction Reconsidered," *Proceedings of The Aristotelian Society*, 104/1 (2003): 1–29.

[2]The sociological study of science started arguably with the French sociologists Émile Durkheim and Marcel Mauss and their interest in how people classify things. The major turning point however came with the studies of Robert K. Merton, especially his book *The sociology of science:*

what we can describe in propositional form. The role of practice therefore becomes central, and with it that of the expert.[3]

Let me now bring the discussion to the frontline of current research and innovation. My focus will be an important recent study on scientific expertise by Harry Collins and Robert Evans, entitled *Rethinking Expertise*.[4] We can describe their overall approach as philosophical sociology because it involves a clarification of the basic conceptual tools needed for the understanding of expertise. In short, their book presents a careful taxonomy of different kinds of expertise, identifies the different questions that scholars need to address, makes a bold proposal about experts who are directly engaged and those who are less engaged, as I will describe shortly, and then offers various case studies of scientific practice as a support for their general thesis. Since this chapter is not meant to describe all this in detail, I will move on directly to the main idea that I think can serve as a useful bridge between scientific and ecclesiastical practices.

This concerns one of the distinctions that Collins and Evans discover between what they call contributory expertise and interactional expertise. For them, we can understand expertise in a very broad sense. Every disposition that a person can acquire is in a way a proto-expertise, or a "ubiquitous expertise", even if this disposition is not spectacular in any way or is not limited to specific individuals. For instance, we can take all Chinese people as experts in speaking Chinese. I consider this very broad sense of expertise of little use, but Collins and Evans have a point in highlighting that, even in this very broad sense, expertise depends on tacit knowledge rather than on logical calculation. To acquire expertise you cannot depend only on the learning of explicit rules. You need enculturation; you need to spend time to absorb the skill via direct physical interaction. Polanyi had famously discussed riding a bike as a case of tacit-knowledge because the physics involved is too complex for those wanting to learn. Some recent critics have challenged this by saying that there is nothing in principle against the possibility of having a large-enough intelligence that can codify and manage all the strict laws of nature involved. Collins and Evans rightly block this objection because cases of tacit knowledge are not designated as

---

*theoretical and empirical investigations*, University of Chicago Press, 1973. Although this was mainly a sociology of the community of scientists, it cleared the ground for a sociological exploration of the cognitive content of science, carried out subsequently by major scholars like Thomas S. Kuhn, Barry Barnes, David Bloor, Gaston Bachelard, Harry Collins, Paul Feyerabend, Steve Fuller, and Bruno Latour. See for instance David Bloor, *Knowledge and social imagery*, London, Routledge, 1976; B. Latour and S. Woolgar, *Laboratory life: The construction of scientific facts*, Princeton University Press, 1986.

[3]In this paper, I focus on science, but the area of philosophy of expertise is much broader. See for instance Evan Selinger & Robert P. Crease, *The Philosophy of Expertise*, New York, Columbia University Press, 2006; Kyle Powys Whyte and Robert P. Crease, "Trust, expertise, and the philosophy of science" *Synthese* 177/3 (2010): 411–425. The question of expertise within ethics has a long history. Some recent scholars have tried to discredit it within this realm of moral philosophy but their case is weak. For a short overview of this debate, see Peter Singer, "Moral experts", *Analysis* 32 (1972): 115-117.

[4]Harry Collins and Robert Evans, *Rethinking Expertise*, University of Chicago Press, 2007.

such simply because of the personal difficulty they would imply had they been just a matter of applying explicit rules. They are designated as such because they involve an added dimension of social-group feedback mechanisms. Think of riding a bike just before the start of lectures in some busy university campus. The skill involves interpreting the intention of others. This additional dimension shows that only social beings like humans can master such skills.[5]

Contributory expertise corresponds to the idea of expertise that we have in our everyday life. Collins and Evans define it as the kind of expertise that "enables those who have acquired it to contribute to the domain to which the expertise pertains" (p. 24). It involves the internalization of physical skills through stages: the stage of the novice, the stage of the beginner, the stage of the competent, and the stage of the expert. The expert in this sense has the ability to do things within that domain, which could be riding a bike, doing original research in the physics of gravitational waves, conducting open-heart surgery, and so on. There is, however, another kind of expertise, a kind that often goes unnoticed. Collins and Evans call it interactional expertise, which corresponds to the kind of expertise that is acquired by those who, although not direct contributors themselves, are so familiar with the contributory experts and with their domain that they come to "know all about it". The basic idea is that you can interact so much and in such fine detail with open-heart surgeons that you will eventually become an expert, even though you never conduct open-heart surgery yourself. Collins and Evans explain: "The idea of interactional expertise implies that the complete fluency in the language of a specialist domain can be obtained in the *absence* [italics in the original] of full-blown physical immersion in the form of life" (p. 30–31). This is the thesis proposed and defended by Collins and Evans that I think can throw light on authority and expertise within the Church. Put simply, the idea is that one acquires *contributory* expertise within a specific domain by doing things; one acquires *interactional* expertise by interacting verbally only. Moreover, Collins and Evans make the bold claim that, from the viewpoint of a third party who seeks advice, the two kinds of expertise are indistinguishable.

## Expertise Within the Church

We move on now to ecclesiastical practice. I need first to express some caution because the question regarding the nature of authority in the Church is obviously vast. There are many distinct kinds of authority we can consider (such as doctrinal,

[5]The point the authors are making is not that only humans are good at interpreting the intention of others. Some non-human animals are, in a sense, perfectly capable of such interpretation and can thus acquire various skills that are comparable to those available to humans. The point is rather that some skills are only possible because humans are capable of interpreting what other humans want, desire or wish. The point here gains plausibility when we recall that human mental states can be highly masked and intricate.

governmental, juridical, and so on), and the debate has a very long history, some of its fundamental ingredients having been part of the teaching of Jesus himself. One way of specifying my focus is to concentrate on the notion of expertise. In the ecclesiastical context, expertise refers to the way the community acknowledges some of its members as having a special kind of knowhow that is useful for the entire community in certain circumstances. The main questions that come to mind here are the following. Who are the experts in the Church? Is there a role for the non-experts in the determination of the correct answer to a particular question? As I mentioned at the beginning, our analysis would be sharper if we make such general and abstract questions more concrete. Hence, the focus question "For policies involving scientific issues, can ecclesiastical decision makers ever be genuine experts when they are not scientists?"

Various theological areas are relevant here. Let me mention briefly the three main ones. First, we have the area that deals with the very nature of authority. In general, authority is a requirement for effective united action within a group. Since humans are political animals, some forms of authority are inevitable even though every form of authority remains in tension with the notion of individual interiority. Ecclesiastical authority follows this general pattern but highlights the divine origin of authority. The church is a society with both a supernatural and a natural dimension. For unified action to be possible, both these dimension require structures of authority. Institutional authority has characterized the Church since its origins but so have charismatic forms of authority. The debate on the nature of expertise can throw new light on these distinctions, especially as regards the way authoritative individuals acquire their status. A second theological area deals with the juridical sense of authority and expertise. As described concisely in Canon 228, "lay persons who excel in necessary knowledge, prudence, and integrity are qualified to assist the pastors of the Church as experts and advisors, even in councils according to the norm of law."[6] The main point regarding authority and expertise here is significant for the analysis I am presenting in this chapter. It highlights the interaction between pastors and experts. The distinction between contributory and interactional expertise is already evident. Before proceeding with this reflection, however, let me mention one last theological area that can be very relevant: the locus of the *sensus fidelium*. The basic idea here is that we should not consider decision-making in the Church an exclusively top-down affair. We should see it rather as a complex dialogical process that includes the hierarchy and also the expertise gained by the laity via their day-to-day direct engagement with the world. Some interpretations of the *sensus fidelium* bring out this dimension very clearly. For instance, Zoltan Alszeghy interprets the *sensus fidelium* as an underlying structured synthesis of lived reality understood in the light of the Gospel. For Karl

[6] *Codex iuris canonici*, Vatican City: Libreria Editrice Vaticana, 1983; the English version, *Code of Canon Law*, is available online: http://www.vatican.va/archive/ENG1104/_INDEX.HTM.

Rahner, it is a precognitive and comprehensive grasp (*Vorgriff*) of reality.[7] Such views highlight the hidden, implicit grasp of a way of life, which gives lay people a kind of fundamental knowhow, a kind of contributory expertise.

Given this background, let us now apply some of the insights we drew from the work of Collins and Evans to the situation of the Church. Practicing scientists are in fact contributory experts. Their day-to-day experience helps them acquire the most fundamental type of expertise that is relevant in this area. The non-scientist clergy who interact seriously and continuously, in a learning mode, with this group of contributory experts can acquire, as Collins and Evans explain, the corresponding interactional expertise. If we follow Collins and Evans all the way, we would even say that the clergy's interactional expertise can eventually extract so much of the essential features of the contributory expertise that the two kinds of expertise will be indistinguishable for a third party. So, regarding the role of expertise, church practice is not out of step with what we find acceptable in scientific practice and in mainstream social structures in general. Those who denigrate the role of experts within the Church are just expressing an unfounded prejudice. Of course, things can go wrong. To attain genuine interactional expertise, the bishops and other clergy concerned need to take interaction seriously. This condition however holds across the entire range of social practice, whether ecclesiastical, scientific or other.

At this point in my argument, before I proceed further with the application of these insights to the case of Vatican Astronomers, I would like to express some reservations about the work of Collins and Evans especially about their bold claim that the two kinds of expertise eventually become indistinguishable. It seems that the justification for such indistinguishability between contributory and interactional expertise depends on a kind of Aristotelian view of practice, whereby a practice is conceivable as matter and from. The intellect of the one who interacts with the contributory expert can come to share, as it were, the formal aspect completely even if he or she does not engage practically. That "matter" of the practice, the practical concrete hands-on experience, can therefore be dispensed with. This is what Collins and Evans assume. In doing so, they disagree with other protagonists in this debate, especially the brothers Hubert and Stuart Dreyfus, who defend the need of a body, a physical body, for the successful acquisition of skills.[8] For the Dreyfus brothers, the direct engagement with the material aspect of the practice gives the upper hand to

---

[7]Z. Alszeghy, "The Sensus Fidei and the Development of Dogma," in: *Vatiĉan II: Assessment and Perspectives Twenty-Five Years after (1962–1987)*, 3 v., ed. R. Latourelle, New York, 1988, vol. 1, pp. 138–156. For Rahner, see "The Faith of the Christian and the Doctrine of the Church," *Theological Investigations* vol. 14, New York, Crossroad, 1976, pp. 24–46; "A Hierarchy of Truths," *Theological Investigations* vol. 21, New York, Crossroad, 1988, pp. 162–167; "The Act of Faith and the Content of Faith," ibid., pp. 151–161; "What the Church officially teaches and what the people actually believe," *Theological Investigations* vol. 22, New York: Crossroad, 1991, pp. 165–175; "The Relation between Theology and Popular Religion," *ibid.*, pp. 140–147. See also O. Rush, "Sensus Fidei: faith 'making sense' of revelation," *Theological Studies* 62 (2001): 231-261.

[8]Stuart E. Dreyfus and Hubert L. Dreyfus, *Mind over Machine: the power of human intuition and expertise in the era of the computer*, New York: Free Press, 1986.

contributory expertise. Collins and Evans try to minimize this aspect, claiming that all we require is a "minimal body", just a brain to think, an ear to hear and a voice to express ourselves. On this crucial point, I tend to sympathize more with the Dreyfus brothers. Dismemberment of humans into body and thought is too crude to be of any good. Moreover, communication is not just language, and social interaction is not just bodily interrelation. Language is continuous with physical behavior. Saying things is a kind of doing things. The idea of meaning is much broader than what we do with sounds.

Given this inevitable merging of matter and thought, of body and language, I am convinced that we cannot accept the equivalence between the two kinds of expertise. This does not mean that interactional expertise is useless or impossible. For instance, I would say that you do not need to be a neurosurgeon to understand the practice of neurosurgery. You do not need to be a full-time astronomer to understand the practice of astronomy. You can become an expert by speaking to many neurosurgeons and many astronomers. Your expertise however will never be the same as that of a neurosurgeon or of an astronomer. It would be an expertise from a different perspective. The best way forward in this debate, therefore, is to use the notion of perspectival appropriation of any given practice. Contributory experts appropriate the skill from one perspective (they are *in* the practice), while interactional experts appropriate it from another perspective (from *outside* the practice). Collins and Evans see the two kinds of expert as engaged in a competition of competence as regards a third party and they argue (referring to the Turing test for artificial intelligence) that the two kinds of expertise are indeed on the same level. It seems to me however that interactional experts are definitely detached from the material side, and this makes them truly deficient when compared to contributory experts. Nevertheless, being detached from the material side offers some benefits. It enables interactional experts to hover above, as it were, the various pockets of contributory experts. It allows them to make comparisons and relative evaluations that are inaccessible to the contributory experts. Interactional experts are freer to move around. Applied to the ecclesiastical situation, this means that although bishops and other clergy can never be experts as regards science in the same way as fulltime scientists, if they interact well enough with scientists, they can acquire genuine interactional expertise. They can acquire a kind of expertise that represents a valuable perspective on scientific practice, a perspective that, in a sense, is unavailable to fulltime scientists themselves.

## The Case of Vatican Astronomers

If we now apply the foregoing insights to the case of Vatican Astronomers, we can highlight at least three important points. First, it should be obvious by now that Vatican Astronomers should be, and indeed are, contributory experts. They have the theoretical foundations, the knowhow, the skills, and the access to professional networks. The mission of the Vatican Observatory would be impossible otherwise.

Secondly, the Vatican Observatory can continue to help the Universal Church by facilitating the acquisition of interactional expertise by ecclesiastical decision makers, at least some of them. If my analysis is correct so far, this means that the Vatican Observatory should allow and even encourage some ecclesiastical decision makers to interact with them regularly. We need to recall that ecclesiastical decision makers are often experts in non-scientific areas and that, for them, the Vatican Observatory would be the only possible contact with the world of fulltime scientists. My third and final point is a word of caution. The contributory expertise represented by the Vatican Observatory is obviously focused on astronomy only. This may not correspond exactly to the real needs of ecclesiastical decision makers who seek an interactional expertise on various fronts of the complex worldview associated with the various sciences. This is a genuine worry. Nevertheless, I am convinced that, although the Vatican Observatory does not have branches in biology, in nuclear physics, in human reproductive technology, and in other significant scientific frontiers, we should still consider it indispensable. Although it cannot offer everything, the amount it can offer is extremely valuable.

The various arguments presented in this chapter are enough, I hope, to show that a study of expertise is indeed useful to explore the notion of authority in various social situations, including scientific and ecclesiastical ones. Of course, there is much more to say about every topic I mentioned. The little I said however seems enough to justify the following two main points, namely that, if we acknowledge the legitimate role of interactional expertise in science, we should also acknowledge it in church practice, and that the Vatican Observatory can be a good source of such interactional expertise to those who want it.[9]

[9]Other works consulted: M. J. Lacey and F. Oakley (eds.), *The crisis of authority in Catholic modernity*, Oxford University Press, 2011; Marie Ann Breitenbeck, *The role of experts in ecclesial decision-making in the 1983 code of canon law*, Ann Arbor MI, University Microfilms International, 1991; Steven Brint, *In an age of experts: the changing role of professionals in politics and public life*, Princeton University Press, 1994. I would like to thank my colleagues of the Vatican Observatory and Prof. Stjepan Marcelja of the Australian National University for their useful comments on a previous draft of this paper.

# Towards Furnishing the Universe

**Paul Mueller, S.J.**

In the century and a quarter since the re-foundation of the Vatican Observatory, there have been many changes in its scientific and cultural context. In this paper I would like to reflect on implications of some of those changes for the mission of the Observatory.

The re-foundation of the Vatican Observatory was confirmed by Pope Leo XIII in the motu proprio *Ut mysticam*, issued on 14 March 1891. Consistent with the leonine spirit of its papal author and its March date of issue, *Ut mysticam* 'comes in like a lion and goes out like a lamb.' The first paragraph is combative in tone: Pope Leo decries the error of those whose "disdain and hatred for the mystical Spouse of Christ" impels them to "calumniate her to unlearned people," calling her "the friend of obscurantism, one who nurtures ignorance, an enemy of science and of progress…"[1] But by the conclusion of the motu proprio the Pope adopts a much gentler tone, seeking to persuade rather than to castigate error. The Observatory will serve "to promote a very noble science which, more than any other human discipline, raises the spirit of mortals to the contemplation of heavenly events"; moreover its re-founding is part of the Pope's plan to help everyone "see clearly that the Church and her Pastors are not opposed to true and solid science, whether human or divine, but that they embrace, it encourage it, and promote it with the fullest possible dedication."[2]

Pope Leo had ample reason to be combative. The *re*-foundation of the Vatican Observatory had been necessitated by, and was in response to, Italy's 1879 expropriation of the papal observatory at the Roman College. And indeed the Church was regularly accused of being a "friend of obscurantism" and an "enemy of science and progress". But then what motivated Pope Leo's shift to a much gentler tone at the conclusion of the motu proprio? One might perceive in this a

[1]Translation from Maffeo S.J. (2001). 315
[2]Ibid. 319.

P. Mueller, S.J. (✉)
Specola Vaticana, Vatican City State, Holy See (Vatican City State)
e-mail: paulmueller.sj@gmail.com

 
G. Gionti, S.J. and J.-B. Kikwaya Eluo, S.J. (eds.), *The Vatican Observatory, Castel Gandolfo: 80th Anniversary Celebration*, Astrophysics and Space Science Proceedings 51, https://doi.org/10.1007/978-3-319-67205-2_19

magnanimous conciliatory gesture on the part of the Pope: from a position of strength he offers an olive branch to the Church's adversaries, providing face-saving 'cover' to any who wish to acknowledge the error of their ways and come home to Mother Church. But I wish to offer a different interpretation.

At the time of the re-foundation of the Vatican Observatory, late in the 19th century, popular faith in the reliability and completeness of science was approaching a zenith. The two previous centuries had witnessed unremitting progress in the development and application of celestial and terrestrial dynamics, with the point of departure being Isaac Newton's law of universal gravitation and laws of motion; and during the immediately previous decades the phenomena of light, electricity and magnetism had been unified and explained, elegantly and powerfully, in James Clerk Maxwell's theory of electro-magnetism. A 1903 remark from physicist Albert Michelson reflects the then-popular conviction that science was approaching a state of completion and perfection: "The more important fundamental laws and facts of physical science have all been discovered, and these are so firmly established that the possibility of their ever being supplanted in consequence of new discoveries is exceedingly remote… our future discoveries must be looked for in the sixth place of decimals."[3]

Even as the cultural power and sway of science was growing dramatically in the latter part of the 19th century, during that same period the Church was falling in popular esteem. The pope was the "prisoner of the Vatican", culturally marginalized and politically enfeebled; the Church was the reputed "enemy of science and of progress," suspect both morally and intellectually. So in 1891, Pope Leo was *not* in a position of strength relative to science and the culture at large; at least in the popular mind, he was in a position of relative weakness. His conciliatory closing remarks in *Ut mysticam* could therefore be seen as supplicatory in nature, even if (as noted above) they were offered under a cloak of magnanimity. The Pope was entreating the ascendant scientific culture to accept and acknowledge the Church as a fellow seeker, friend, and guarantor of truth. The mission of the re-founded Observatory would be to conduct scientific research in such a way as to satisfy the standards of the ascendant scientific culture—and to do so in service of Pope Leo's apologetic agenda, showing that "the Church and her Pastors are not opposed to true and solid science, whether human or divine, but that they embrace it, encourage it, and promote it with the fullest possible dedication."[4] In light of the apologetic aspect of its stated mission, the Vatican Observatory was re-founded in something of a defensive crouch. It was tasked with doing good science, not just in order to contribute to human knowledge but also to plead in the High Court of Science on behalf of the Church's *bona fides* as seeker, friend, and guarantor of truth.

It remains the case now, a century and a quarter later, that the mission of the Vatican Observatory goes beyond just doing good science—the Observatory is still

[3]Michelson (1903). Paraphrased versions of this remark have often been attributed erroneously to Lord Kelvin.

[4]Maffeo S.J. (2001). 319.

tasked with doing "something more". But today that "something more" has less to do with pleading on behalf of the Church in the High Court of Science, and more to do with representing *both* the Church *and* Science in the High Court of the Culture-at-Large, where the very notion of scientific truth is on trial. Far from being in the position of having something to prove to the world of science, nowadays the Church and Vatican Observatory are making common cause with science in defense of scientific truth. Allow me to elaborate:

Since the start of 20th century, and especially in the post-Conciliar period, the Church's perspective on modern science and scientific reasoning has been one of deepening sincere interest. Pope John Paul II, drawing on long-standing Church doctrine and tradition, saw faith and reason not as competitors or enemies but as two aspects of the search for truth: "What is distinctive in the biblical text is the conviction that there is a profound and indissoluble unity between the knowledge of reason and the knowledge of faith. The world and all that happens within it, including history and the fate of peoples, are realities to be observed, analyzed and assessed with all the resources of reason, but without faith ever being foreign to the process… There is thus no reason for competition of any kind between reason and faith: each contains the other, and each has its own scope for action."[5] And more: "The truth, which God reveals to us in Jesus Christ, is not opposed to the truths which philosophy perceives. On the contrary, the two modes of knowledge lead to truth in all its fullness. The unity of truth is a fundamental premise of human reasoning…"[6] According to John Paul II, truth is revealed in the 'two books' of Nature and Scripture. Since these two books are written by one and the same author, they cannot disagree with each other: truth cannot contradict truth. Thus any apparent disagreement between the 'two books' must derive from limitations of human understanding, and is an indication that we have not yet arrived at an accurate reading of one or the other or both of the two books. In such situations the appropriate pastoral response is to offer consolation and to counsel forbearance and patience; the appropriate intellectual response is to do more and deeper research.

Pope Benedict XVI offered a nuanced appreciation of the respective competencies of science and faith. The great and particular glory of physical science is its capacity to show cumulative progress, advancing towards truth over time: "… incremental progress is possible only in the material sphere. Here, amid our growing knowledge of the structure of matter and in the light of ever more advanced inventions, we clearly see continuous progress towards an ever greater mastery of nature."[7] But the same cannot be said of the realm of ethics, morals, and faith, where

[5]Pope John Paul II. Encyclical Letter *Fides et Ratio*, promulgated September 14, 1998. #16–17. See http://w2.vatican.va/content/john-paul-ii/en/encyclicals/documents/hf_jp-ii_enc_14091998_fides-et-ratio.html.

[6]Ibid. #34.

[7]Pope Benedict XVI. Encyclical Letter *Spe Salvi*, promulgated November 30, 2007. #24. See http://w2.vatican.va/content/benedict-xvi/en/encyclicals/documents/hf_ben-xvi_enc_20071130_spe-salvi.html.

> …there is no similar possibility of accumulation for the simple reason that man's freedom is always new and he must always make his decisions anew… Freedom presupposes that in fundamental decisions, every person and every generation is a new beginning. Naturally, new generations can build on the knowledge and experience of those who went before, and they can draw upon the moral treasury of the whole of humanity. But they can also reject it, because it can never be self-evident in the same way as material inventions. The moral treasury of humanity is not readily at hand like tools that we use; it is present as an appeal to freedom and a possibility for it."[8]

Cumulative progress is possible in the physical sciences, where as Isaac Newton noted we "stand on the shoulders of giants", with yesterday's cutting-edge discovery regularly becoming tomorrow's taken-for-granted tool.[9] But similar cumulative progress would be possible in the realm of ethics and morals only if there were no real human freedom. If we are truly free, then in ethics and morals we can consult the "giants" but we cannot stand on their shoulders. And according to Benedict this should be considered a *feature* of ethics and morals—not a *bug*!

In summary, since the start of the 20th century and especially during the post-conciliar period the Church has grown more open, confident and comfortable in its rapport with modern science. Yet during that same period popular faith in the reliability of science as guide to truth has been on the wane—it is no longer anywhere near the zenith it reached at the turn of the 20th century. This is for several reasons. First, a lesson learned from the shock of the great 20th century revolutions in physics (relativity and quantum theory) is that one should never presume that science is approaching a state of perfection and completion: even the dearest-held scientific theory can be overthrown. Reinforcing that lesson is the uncomfortable fact that the currently-accepted versions of relativity and quantum theory are logically incompatible: though both theories make fantastically accurate predictions, and both seem indispensable, they cannot both be true—at least not in their current forms. Moreover, the counter-intuitive strangeness of quantum theory led it to be understood via the "Copenhagen interpretation", which jettisons the notion of a thoroughgoing correspondence between theory and reality: various aspects of the mathematical machinery of quantum theory are interpreted not as corresponding to something "in reality", but rather as being tools or instruments that are useful for calculating the expected results of observations and measurements. Quantum theory is the most accurate and powerful of scientific theories, but its counter-intuitive strangeness has served to problematize what it even means to hold that a scientific theory is *true*.

Second, the concerted effort over the last century to determine the true and reliable method of science—and in particular the true and reliable method of determining which of various rival scientific theories is true—has created more heat than light. Early 20th century philosophers presumed that observational data alone could serve to adjudicate objectively between competing theories. But later in the

[8]Ibid.

[9]"If I have seen further it is by standing on the shoulders of giants." Isaac Newton, Letter to Robert Hooke, 15 February 1676.

century, Norwood Russell Hanson and Thomas Kuhn argued convincingly that observational data cannot be cleanly separated from theory, and thus the theoretical context from within which scientific theories are evaluated can influence that very evaluation—with the result that scientific observation is incapable in principle of distinguishing with complete objectivity between competing scientific theories.[10] Still later in the 20th century, a series of seminal works explored the historical contingency of methods and criteria of theory-choice in science.[11] The resultant insight was that there is no such thing as Science with a capital S, unique and universal. This is not to deny that the sciences move progressively towards truth—but it does challenge the notion of a unique and universally valid method in science.

Third, subsequent scholars initiated a variety of sociological-cultural-anthropological deconstructions of the perceived power and hegemony of science. For example, it was claimed in some quarters that supposedly true theories of empirical science are "merely" social constructs, or are "merely" culturally determined. On this reading, the replacement of an older theory by a newer one should be explained not by the new theory's greater "truth" but rather in terms of sociological power relationships, or in terms of cultural context, or in terms of the anthropological practices of the tribe of scientists. These scholars rejected the notion that science should be accorded some special status or respect because its theories "correspond" to something real about the natural world. These (somewhat arcane) academic disputes and claims erupted into the public eye in the notorious "Science Wars" of the early 2000s. On one side were those who sought to challenge the prestige and dominance of the physical sciences by claiming that theory acceptance in science has little or nothing to do with truth, and is really just a matter of social convention, political power, or post-modern posturing and performance. On the other side were scientists and others who were outraged or bemused by such claims. The issue came to a head when physicist Alan Sokal succeeded in publishing an article in the journal *Social Text* in which he argued at that quantum gravity is a linguistic and social construct and that quantum physics supports postmodern criticisms of scientific objectivity.[12] Soon thereafter Sokal revealed, in an article published elsewhere, that his *Social Text* article had been an intentional parody—a tongue-in-cheek experimental test of the intellectual rigor of the journal *Social Text.*[13] In Sokal's view, his "experiment" revealed the lack of academic rigor of a journal that that would "publish an article liberally salted with nonsense, if it sounded good and if it flattered the editors' ideological preconceptions".[14] The editors of *Social Text* charged Sokal with acting in bad faith, saying that, as editors of an interdisciplinary journal, they have no choice but to rely upon the professional

[10]Hanson (1958), Kuhn (1962).

[11]See for example Hacking (1983), Shapin and Schaffer (1985), Gooding et al. (1989), Cartwright (1983), Galison (1987).

[12]Sokal (1996a).

[13]Sokal (1996b).

[14]Ibid.

reputation and honesty of contributors from diverse fields. The “Sokal Affair” became a *cause célèbre.* Sokal’s intention was to undergird the notion of objective truth in science. Though he succeeded in persuading the already-converted, in many cases the effect of the “Sokal Affair” was to reinforce a perception that science, and in particular theory-choice in science, is more about tribalism and power than it is about objective truth.

To sum up: In the 21st century, science plays a peculiar mixed role in the public consciousness. Viewed as a source of technology—and especially as a source of new consumer technologies—science is held in the highest esteem. In much of the world, such technologies have become so deeply enmeshed in (and even constitutive of) everyday life that it is difficult to imagine getting along without them. Insofar as we depend on the technology embedded in smartphones and computers, we acknowledge at least tacitly the reliability of the science behind that technology—and to that extent science is held in higher esteem than ever. On the other hand, following upon the above-mentioned 20th century disputes over scientific theory-choice, and perhaps also due to resurgent fundamentalism and to the tribalization and Balkanization fostered by the echo chamber of social media, it has become increasingly difficult in recent decades to engage in reasoned public discourse concerning (for example) the reality and causes of biological evolution and climate change, or the safety of vaccination and water fluoridation. In these areas and in many others, the authority of scientific expertise is regularly dismissed as suspect by large segments of the public—and to that extent science is not held in high esteem.

This is the context of the Vatican Observatory’s mission today. It seems clear that the modern mission of the Observatory must include a response to that context. Certainly the Observatory’s mission remains that of doing science, in collaboration with and according to the standards of the scientific academy. But in today’s context, the Observatory is no longer in the position of needing to prove, to an ascendant scientific academy, that it and the Church really are seekers and friends of truth. Instead the Observatory’s mission must now include making common cause with the (perhaps somewhat chastened) scientific academy in rehabilitating the very notion of scientific truth and concomitant expertise. One might think it useful and relevant for the Observatory’s modern mission that its Jesuit staff, unlike most professional scientists, has been trained in philosophy and theology. Perhaps these Jesuit philosopher-scientists will be able to offer insights and arguments that are apt and persuasive. But then again, maybe not. In the modern context, philosophical and theological argument seem to have little effect on the public consciousness; as people have come to reject the authority of expertise, they have become correspondingly immune to the power of reasoned argument. What ends up being determinative, all too often, is not reasoned argument but tribal identity.

I would like to speculate concerning a general framework and strategy for responding to the modern context. What is needed is a rehabilitation of the very notion of scientific expertise—not so much expertise in theory-choice, but rather expertise in observation. Everyone is familiar with habit, and it is habit that makes observation possible. If you didn’t form the *habit* of seeing the same object as being the same size, regardless of how far away it is from you, it would be difficult for you to say what you mean by calling it the "same" object. And so I propose: if people could be led, via a series of pastoral/intellectual exercises or practices, to become more self-aware concerning the prominent role of habit in the ordinary perception on which we rely every day, perhaps they could be led to see, appreciate, and trust the even deeper role that habit plays in expert perception, as practiced by scientists. Expert scientific observation *is* a kind of habit—a refined habit, but a habit nonetheless. People rely on others' habits often in the practical order of daily life. Why then should they not be willing to rely on others' more refined habits—the habits that constitute expertise in observation?

I follow my (much esteemed!) past professor Lorraine Daston in holding that there is no one who gets at this issue more clearly than Polish bacteriologist Ludwig Fleck.[15] Speaking about microscopic observation of bacteria, Fleck writes: "Direct experience of form requires being experienced in the relevant field of thought. The ability directly to perceive meaning, form, and self-contained unity is acquired only after much experience, perhaps with preliminary training. At the same time, of course, we lose the ability to see something that contradicts the form. But it is just this readiness for directed perception that is the main constituent of thought style."[16] A novice can see only blurs and blobs under the microscope; experience and training are required in order to make sense of the visual chaos—in order to be able to see *things*.

For Fleck, learning to see and observe like a scientist is a matter of accumulated experience—not just of an individual, but of a well-trained collective. It is through virtuoso perception that we succeed in forging stable kinds out of confused sensations. Fleck sounds similar to Aristotle, who wrote:

> So from perception there comes memory, as we call it, and from memory (when it occurs often in connection with the same thing), experience; for memories that are many in number form a single experience. And from experience, or from the whole universal that has come to rest in the soul (the one apart from the many, whatever is one and the same in all those things), there comes a principle of skill and of understanding.[17]

Without acquired habits of perception, which are cultivated by observation, there would be no science, and no sensible world at all. Perception doesn't *create* the universe, but it does *furnish* it: shaping and sorting, arranging parts into wholes, distinguishing one thing from another. I recommend Ludwig Fleck's analysis of expert observation to the Jesuits of the Vatican Observatory, as a possible aid in coming to greater awareness of the nature of their own expert observation, and of the role of habit therein. Informed by Fleck, the Observatory Jesuits might find themselves better positioned to persuade others to recognize the role of habit in

[15]Daston (2008). See page 99ff.

[16]Fleck (1979). 92.

[17]*Posterior Analytics* 2.19.100a4-8.

everyday observation. That could help open the door to seeing and trusting various kinds of expertise. And that could in turn open the door to a popular rehabilitation of scientific truth.

There are recognized virtuoso practitioners and observers not only in science but also in the Christian life. There are not many who have observed the movements of the interior life and soul with the virtuosity of Augustine, or John of the Cross, or Ignatius Loyola. There are not many who have succeeded in perceiving relationships between God, humanity and world with the virtuosity of Aquinas, or Newman, or Rahner. There are not many who have plumbed the beauty and depth of Christian symbol and imagery with the virtuosity of Michelangelo, or Caravaggio, or Donne, or Hopkins. If virtuoso scientific observers have provided us with 'technologies' for manipulating and controlling the world, virtuoso Christian observers have provided us with 'technologies' for thriving, living, and loving—technologies that have endured for 2000 years through many grave setbacks. We are all supported, in both spiritual and physical life, by the accomplishments of virtuoso observers.

In short, I propose *virtuosity* as a point of contact and commonality between science and faith, one quite relevant to the contemporary situation. Science and faith both rely on virtuoso observation to furnish the universe with objects and things from which practical 'technologies' eventually 'trickle down' to sustain, enlighten, and facilitate the lives of those who are not virtuosi. The long-term success of such trickle-down technologies serves as a kind of confirmation that the way the world has been furnished by the virtuosi is more or less correct.

I will not presume to claim that the Jesuits of the Vatican Observatory are geniuses on the order of the names listed above! But they have a rare gift, in that they have been trained to be virtuosi not only in science but also in spirituality and prayer. This poses a danger. Virtuosi can see things that regular people cannot: perhaps patterns in stellar spectra or membership in star clusters, or perhaps or movements of the spirit and signs of spiritual desolation. Also virtuosi no longer see things that other people continue to see—virtuosi automatically "edit out" of their perception that which they have learned to interpret as mere noise or artifacts. Accordingly there is a danger that the Jesuits of the Observatory, as virtuosi, will talk past "regular people" and that their expertise will be dismissed as irrelevant. A special effort is required, one both intellectual and pastoral, for the Observatory Jesuits to carry out the mission of communicating effectively with non-experts.

But if there is a special danger, there is also a special opportunity. Jesuit training has the effect of making Jesuits rather self-conscious and self-reflective concerning the process of their own formation: they reflect upon the nature of their own training and formation in a way in which scientists typically do not. I suggest that the Jesuits of the Observatory could constitute, among themselves, a kind of informal laboratory for the study and exploration of expert/virtuoso observation. They have been formed to be virtuosi, not only in science but also in spirituality and prayer. Perhaps from that peculiar expertise will flow pastoral and intellectual wisdom concerning the popular rehabilitation of scientific expertise and truth.

## References

Cartwright, N. (1983). *How the laws of physics lie*. Oxford: Oxford University Press.
Daston, L. (2008). On scientific observation. *Isis, 99*(1), 97–110.
Fleck, L. (1979). *Genesis and development of a scientific fact*. Chicago: University of Chicago Press.
Galison, P. L. (1987). *How experiments end*. Chicago: University of Chicago Press.
Gooding, D., Pinch, T., & Schaffer, S. (1989). *The uses of experiment: Studies in the natural sciences*. Cambridge: Cambridge University Press.
Hacking, I. (1983). *Representing and intervening: Introductory topics in the philosophy of natural science*. Cambridge: Cambridge University Press.
Hanson, N. R. (1958). *Patterns of discovery: An inquiry into the conceptual foundations of science*. Cambridge: University Press.
Kuhn, T. S. (1962). *The structure of scientific revolutions*. Chicago: University of Chicago Press.
Michelson, A. A. (1903). *Light waves and their uses* (Vol. 3 of the 2nd series). Chicago: The University of Chicago Press.
Maffeo S.J., S. (2001). *The Vatican observatory: In the service of nine popes*. Vatican Observatory Publications.
Shapin, S., & Schaffer, S. (1985). *Leviathan and the air-pump: Hobbes, boyle, and the experimental life*. Princeton: Princeton University Press.
Sokal, A. D. (1996a). Transgressing the boundaries: Towards a transformative hermeneutics of quantum gravity. *Social Text, 46*(47), 217–252.
Sokal, A. D. (1996b). A physicist experiments with cultural studies. *Lingua Franca*, 62–64.

# Conclusions

The Specola Vaticana (Vatican Observatory) was founded, as it is written in its founding document, the *Motu Proprio* "*Ut Mysticam*" of Pope Leo XII in 1891, to show to the world that the Catholic Church is not opposed to robust and solid science. This goal was thought to be achieved by having staff scientists who were (and are) consecrated people (either religious brothers or religious priests, though in principle religious women could also be employed at the Vatican Observatory) who are doing science, to show to the world that scientific research and religious beliefs are really not incompatible. Now after 126 years from the re-foundation of the Vatican Observatory and 82 years from the inauguration of the Vatican Observatory in Castel Gandolfo, one could ask if this goal of showing to the world that science and the Catholic faith are totally compatible has been reached.

First of all, it is worth noticing that the "climate" is a bit changed respect to 1891, that time of the modernist attack and the rising of the positivism in the culture. Now we live in a "postmodern" era in which the role of science in the cultural world does not have, anymore, a privileged status as "queen of knowledge," producing a knowledge that is "absolutely" true. Instead, the postmodern critique has attacked science, promoting the idea that science, especially western science, is like any human discipline a "social" construct and so produces knowledge that is simply a human construct, valid mainly for the western world. There are examples of sciences developed in other cultures, like eastern medicine for example, which were useful for that culture, and are different from western culture. Although this critique contains some truth in it, it is a bit extreme. Western science contains some "universality" in it.

Meanwhile, the axis of the fight between science and faith has shifted from the purely academic arena to a wider contemporary culture, especially in the west, in which a broader mass education has implied that people with average culture have, today, a better sense than in the past of the main themes debated in science. Knowing science at, at most, only a popular level, these people may be more influenced by superficial arguments against religion from exponents of the movement called "the new atheism".

How has the Observatory responded to this change? The papers presented at this 80th Anniversary Conference have shown that the way of doing research of the



G. Gionti, S.J. and J.-B. Kikwaya Eluo, S.J. (eds.), *The Vatican Observatory, Castel Gandolfo: 80th Anniversary Celebration*, Astrophysics and Space Science Proceedings 51, https://doi.org/10.1007/978-3-319-67205-2

Vatican scientists has also completely changed compared to the early stages of the newly refounded Observatory in 1891. Vatican astronomers today work with other lay colleagues in common research projects and not, as in the past, in an isolated tower far from the world. This fact has a double positive effect. Vatican scientists are supported and acknowledged by their lay colleagues, who benefit from their expertise and so, indirectly, recognize their scientific authority. Furthermore, being integrated within their particular scientific communities, they become the pastor of this community in which their colleagues are more willing to talk with them about religious issues (even if they are atheists or agnostic) and, if believers, share their faith with them. Meanwhile, Vatican astronomers show by the mere fact of working in science and being either consecrated religious or priests that the majority of the arguments promoted by new atheism movements are simply superficial and, in the majority of cases, nonsense.

We can expect that the near future of the Vatican Observatory will continue along the lines already mentioned. The main work of Vatican scientists will continue in projects with external collaborators to the Observatory.

In the field of *Planetary Science*, the group will continue to work in the direction already marked in previous years. Given that the Vatican Observatory posses a large meteorite collection, it will continue to measure the physical characteristics of meteorites with the aim to obtain results useful to understanding the formation and evolution of the Solar system, the Giant Planets etc. There is also an interesting project, already underway, to detect fireball trajectories using cameras mounted in several different locations around Tucson and thus have a better sense of where those meteorites come from. Another long term project in this field is to determine the physical characteristics of Near Earth Objects (NEO). This is carried out by doing spectrographical analysis of the NEOs using the spectrometer connected to the Vatican Advanced Technology Telescope (VATT), which is called VATTspec. As in the case of the meteorites, this research is useful in the study of the origin and evolution of the Solar System. Extrasolar planets will also continue to be a subject of research in the near future as well.

Research on *Stars and Stellar Evolution* at the Vatican Observatory will continue to see the use of the VATT telescope. There will also be, as now, computer N-Body simulation of the evolution of stars like type B subdwarf stars. Research on stars will continue with the spectroscopic analysis and classification of $\lambda$-Bootis stars and the photometry analysis of nearby stars using the Vilnius Photometric System. Asteroseismology carried out using VATTSpec may be also a subject of future research.

The study of *Galaxies* at the Vatican Observatory will continue to be pursued studying the Widefield Nearby Galaxy Cluster Survey (WINGS) in which it will be possible to also get information on star evolution and the behavior of our galaxy as well, and thus constrain the $\Lambda$CDM cosmological model. Another active field of research in this field will continue to be the evolution, the stellar distribution, and formation of our Milky Way Galaxy as well as other galaxies. In this direction, in the near future some new directions may be the numerical study of the evolution of galaxies. There may also be the study of galaxy halos to detect the presence of Dark Matter.

*Classical and Quantum Cosmology* will continue to be a field of research. Following the death of Fr. Bill Stoeger, S.J., the interest will shift more to the Quantum Gravity and Quantum Cosmology side. Here the research in the near future will be on the early universe, in the sub-Planckian era, as well as aspect of the physics of the beginning of the universe (String Theory, Modified Gravity, Asymptotic Safety). As new scientific personnel join the Observatory, one possible new field of investigation might be in the study of the possible constraints which different theories of Quantum Gravity impose on the Cosmic Microwave Background Radiation (CMB) as well as the study of the same CMB radiation from a theoretical as well as observational point of view.

Finally, *Public Outreach* will always be an important field of activity of the Vatican Observatory. This includes both popular science talks as well as science-theology talks. This is certainly an important, although not primary, part of the mission of the Observatory, "to show to the world that the Church is doing solid and robust science" ("*Ut Mysticam*"...). But this aspect of the mission cannot be done or justified if the members of the Observatory do not do sound scientific work. In fact, only their competencies in their fields of research can give them the tools to explain their subjects in a popular way, communicating to non-experts the key concepts of their research. Vatican Observatory scientists generally have a broader education than their scientist colleagues, as their religious training means that they have also undergone philosophical and theological studies. This additional education, as well as their spiritual training, enables them to bring unique skills to their popular talks. This makes them the right persons to think and deliver talks in science and theology, since they live this tension between Science and theology in their everyday life.

The manufacturer's authorised representative in the EU is Springer Nature Customer Service Centre GmbH, Europaplatz 3, 69115 Heidelberg, Germany. If you have any concerns regarding our products, please contact ProductSafety@springernature.com

Printed and bound by CPI Group (UK) Ltd, Croydon, CR0 4YY
01/10/2026
02225973-0005